EPA/625/R-02/009
www.epa.gov/empact
December 2002

Environmental Curricula Handbook: Tools in Your Schools

National Risk Management Research Laboratory
Office of Research and Development
U.S. Environmental Protection Agency
Cincinnati, Ohio 45268

Office of Environmental Information
U.S. Environmental Protection Agency
Washington, DC 20460

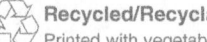

Recycled/Recyclable
Printed with vegetable-based ink on paper that contains a minimum of
50% postconsumer fiber content processed chlorine-free.

Disclaimer

This document has been reviewed by the U.S. Environmental Protection Agency (EPA) and approved for publication.

Acknowledgments

The development of this handbook was managed by Dr. Dan Petersen (U.S. Environmental Protection Agency). While developing this handbook, we sought the input of many individuals. Gratitude is expressed to each person for their involvement and contributions.

Beth Gorman, Pima County Department of Environmental Quality, Tucson, AZ

Susan Green, Northeast States for Coordinated Air Use Management (NESCAUM), Boston, MA

George Host, University of Minnesota, Natural Resources Research Institute, Duluth, MN

Kristin Kenausis, U.S. Environmental Protection Agency, Washington, DC

Richard List, Syracuse City School District, Syracuse, NY

Kim Ornberg, Seminole County Public Works Department, Stormwater Division, Sanford, FL

Curry Rosato, City of Boulder Public Works/Utilities, Water Quality and Environmental Services, Boulder, CO

Julie Silverman and Kara Lenorovitz, Center for Lake Champlain, Burlington, VT

Jodi Sugarman-Brozan, Alternatives for Community and Environment, Roxbury, MA

Pete Tebeau, University of Connecticut, Bridgeport, CT

Rudi Thompson, University of North Texas, Dallas, TX

John White, U.S. Environmental Protection Agency, Research Triangle Park, NC

Adam Zeller, Earth Day Coalition, Cleveland, OH

Contents

1.0 Introduction

Environmental education is a learning process that increases people's knowledge and awareness about the environment and associated challenges, develops the necessary skills and expertise to address the challenges, and fosters attitudes, motivations, and commitments to make informed decisions and take responsible action (UNESCO, Tbilisi Declaration, 1978).

1.1 What Was EMPACT?

The U.S. Environmental Protection Agency (EPA) created the Environmental Monitoring for Public Access and Community Tracking (EMPACT) program to take advantage of new technologies that make it possible to provide environmental information to the public in near real-time. EPA partnered with the National Oceanic and Atmospheric Administration (NOAA) and the U.S. Geological Survey (USGS) to help achieve nationwide consistency in measuring environmental data, managing the information, and delivering it to the public. Through the use of grants, EMPACT helped local governments build monitoring infrastructure in metropolitan areas across the country, addressing questions such as:

- What is the ozone level in my city today?

- How is the water quality at the beach today?

- What is the UV Index in my area today?

EMPACT projects aim to help communities:

- Collect, manage, and distribute time-relevant environmental information.

- Provide their residents with easy-to-understand, practical information they can use to make informed, day-to-day decisions.

Some projects were initiated directly by EPA; others were launched by communities with the help of EPA-funded "Metro Grants." EMPACT projects helped local governments build monitoring infrastructures and disseminate environmental information to millions of people.

EMPACT projects have been initiated in 156 metropolitan areas. These projects cover a wide range of environmental issues, such as groundwater contamination, ocean pollution, smog, and overall ecosystem quality. Having met the program goals, EMPACT ended in 2001. Many projects continue to provide realtime environmental information to local residents.

Recognizing that educating our youth is vital to the future of our planet, many EMPACT projects have incorporated curricula- or school-based components. The curricula are hands-on in their approach and complement the objectives of their associated EMPACT projects. Therefore, the activities and lessons either involve the utilization of monitoring data collected under a particular project or encourage student monitoring to assist project efforts.

1.2 What Is the Purpose of This Handbook?

This handbook is designed to provide teachers and other educators with guidance on how to teach students about environmental issues related to air, water, and soil quality. It provides information to help educators incorporate environmental education into the classroom. The handbook is organized as follows:

- **Chapter 2: How Do EMPACT Programs Work in Schools** discusses why environmental education is important, how to incorporate the lessons and ideas highlighted in this handbook into age-appropriate curricula, and how to identify quality environmental education materials.

- **Chapter 3: Teaching the Teacher–How Do I Make an EMPACT on My Students?** provides background information on air, water, and soil and why we should be concerned about the quality of these substances.

- **Chapter 4: Air-Based Projects** covers the air-based EMPACT projects and their curriculum components.

- **Chapter 5: Water-Based Projects** covers the water-based EMPACT projects and their curriculum components.

- **Chapter 6: Land-Use and Soil-Based Projects** covers the land- and soil-based EMPACT projects and their curriculum components.

This handbook can assist educators in designing lesson plans and activities to teach the principles of environmental science. It highlights a host of EMPACT projects that have developed or are developing curricula or other classroom materials to foster student learning. The highlighted projects cover a variety of grade levels (see Appendix C: Activities by Grade Level). Therefore, this handbook can be used by any teacher, from kindergarten through grade 12. In addition, college-level materials have been developed for some projects. Moreover, in most cases, the activities and lessons geared towards one particular grade can easily be adapted for others. Teachers and educators can review the project descriptions and read about the activities, lesson plans, and tools they employ to develop ideas for their own classrooms. In addition, the handbook includes resources and contact information and in some cases a Web site where lesson plans and activities can be accessed directly.

This handbook also references supplementary sources of information, such as Web sites, publications, organizations, and contacts, that can help the user find more detailed guidance. (See Appendix A: Additional Resources)

2.0 How Do EMPACT Programs Work in Schools?

2.1 Environmental Education—Why Teach Students About the Environment?

Environmental information is important because it affects our daily lives. For example, if you know the air quality is poor on a particular day, you might choose to skip your daily jog or exercise early in the morning when air quality is usually better. Environmental education typically incorporates aspects of economics, culture, politics, and social equity, as well as natural processes and systems. Teaching young people about the environment can help them see the many ways in which people affect the world around them by their actions today, which have consequences for the future health of the environment.

Environmental education can foster in children of all ages an awareness and sensitivity to the natural world, inspiring students to increase their knowledge of the environment, identify environmental challenges, and become motivated about resolving these challenges.

Learning about environmental challenges can also show students first-hand how their individual and collective actions can affect their own health, the environment, the country, and society as a whole. As a result, learning about the environment can help young people make informed day-to-day decisions, influence their peers and caregivers, and grow up to be better citizens.

2.2 Lesson Creation 101— How to Incorporate EMPACT Lessons and Ideas Into Age-Appropriate Curricula

The EMPACT tools described in this handbook use real-time technologies to help develop children's research and reasoning skills. Lessons focus on inquiry-based, hands-on learning. Students not only learn about environmental issues but also are encouraged to explore how feelings, experiences, attitudes, and perceptions influence these issues. This type of teaching helps students develop

Reducing the Risks

Children can be exposed to a number of environmental hazards in their homes, schools, and playgrounds—from tobacco smoke to lead-based paint. Environmental education can help raise teacher, parent, and student awareness of these risks, thereby helping to reduce children's exposure to these hazards over time.

For example, asthma is currently the most common chronic childhood illness in the United States. Over the past 15 years, major advances have been made in understanding the complex interplay between asthma, environmental exposures, and other factors.

This knowledge is helping pediatricians, schools, children, and their caregivers take steps to not only mitigate asthma triggers, but also to learn how to manage this illness on a day-to-day basis (i.e., on high ozone days, asthmatics should not play outside).

the critical-thinking, problem-solving, and team-working skills needed in today's technology- driven world.

EMPACT lessons typically use hands-on, laboratory-based approaches, such as those favored by groups like the National Science Teachers Association (NSTA) and the National Science Foundation (NSF). As such, they often fit best in a science curriculum, but they are also often multidisciplinary, so that the lessons can be incorporated into many different subject areas.

While science forms the foundation for many of the EMPACT lessons in this handbook, social science, health, language arts, math, and other subjects are also covered, as they are critical to fully understanding environmental issues and their impacts on society. (See Appendix D: Activities by Subject.)

For example, the Northeast Ohio (NEO) EMPACT project teaches students about air quality and urban sprawl through a set of 10 hands-on exercises and science experiments. Also included in the lessons are activities that develop language arts skills, such as composing a letter about acid rain for local legislators or completing air quality word searches and crossword puzzles.

The tools referenced in this handbook also serve a range of ages and grades. EMPACT lessons at the primary grades are designed so that younger children can explore the environment and learn basic concepts. At the higher grades, children perform increasingly more sophisticated experiments and data gathering and interpreting tasks.

For example, in the ECOPLEX curriculum (K-8), kindergartners take ultraviolet-sensitive beads outside to see how the beads change colors, thereby discovering where and when the sun's ultraviolet rays are strongest. At the third grade level, students use construction paper and colored pattern blocks to learn how oxygen is converted to ozone. Eighth graders learn how chlorofluorocarbons (CFCs) contribute to ozone depletion through chemistry experiments that demonstrate how compounds separate in a chemical reaction.

A number of the EMPACT tools described in this handbook teach global issues via a local or regional environmental problem; others have a national scope, and some projects reinforce the national scope by enabling students to exchange data and observations with other classrooms across the country.

Finally, most EMPACT lessons have been developed with the help of both technical and curriculum experts, ensuring their accuracy and applicability to state and national education standards.

2.3 Making the Grade—How to Identify and Use Quality Environmental Education Materials

EMPACT tools, like all quality environmental education materials, encourage exploration. Acquiring information changes from a static to active learning process. Students participate in defining goals, gaining knowledge, and presenting results in a variety of formats.

How can schools recognize and use quality environmental education materials? According to the North American Association for Environmental Education

(NAAEE), quality environmental education materials should possess six key characteristics, as listed below. It is useful for educators to be aware of these characteristics and to reinforce them in the classroom when teaching students about the environment.

#1 Fairness and accuracy. Environmental education materials should be fair and accurate in describing environmental problems, issues, and conditions, and in reflecting the diversity of perspectives on them. Materials should have factual accuracy, a balanced presentation of differing viewpoints and theories, openness to inquiry, and reflection of diversity.

#2 Depth. Environmental education materials should foster awareness of the natural and built environment, an understanding of environmental concepts, conditions, and issues, and an awareness of the feelings, values, attitudes, and perceptions at the heart of environmental issues, as appropriate for different developmental levels. Materials should focus on concepts that are set in a context that includes social and economic as well as ecological aspects and demonstrate attention to different scales.

#3 Emphasis on skills building. Environmental education materials should build lifelong skills that enable learners to prevent and address environmental issues. Materials should encourage the use of critical thinking and creative skills. Students should learn to arrive at conclusions about what needs to be done based on thorough research and study and should gain basic skills to participate in resolving environmental issues.

#4 Action orientation. Environmental education materials should promote civic responsibility, encouraging learners to use their knowledge, personal skills, and assessments of environmental issues as a basis for environmental problem solving and action. Materials should instill a sense of personal stake, responsibility, and self-efficacy.

#5 Instructional soundness. Environmental education materials should rely on instructional techniques that create an effective learning environment. Instruction should be learner-centered—materials should offer different ways of learning, and there should be a connection to everyday life. In addition, learning should occur in environments that extend beyond the boundaries of the classroom, and materials should recognize the disciplinary nature of environmental education. The goals and objectives of the materials should be clear, the materials should be appropriate for specific learning settings, and they should include a means for assessing learner progress.

#6 Usability. Environmental education materials should be well designed and easy to use. Materials should be clear and logical to both educators and learners, inviting and easy to use, long-lived, adaptable, and accompanied by instruction and support. In addition, materials should make substantiated claims and fit in with national, state, or local requirements.

For more information on NAAEE's *Environmental Education Materials: Guidelines for Excellence*, visit <www.naaee.org>.

3.0 Teaching the Teacher: How Do I Make an EMPACT on My Students?

3.1 Air

Why should we be concerned about air quality?

Air quality in many U.S. cities is being degraded by human activities such as driving, chemical manufacturing, the burning of fossil fuels, and other industrial and commercial operations. Air pollution also comes from smaller, everyday activities such as dry cleaning or filling your car with gas. As more people drive vehicles, require more electricity, and conduct other activities, more gases and particles are added to the air we breathe. This pollution can reach levels dangerous to humans and the environment.

While air pollution poses a health risk to all humans, it is especially dangerous for children and people with respiratory illnesses. The biggest air pollution-related health threat to children is asthma. Other problems associated with high levels of air pollutants, such as ozone, include irritated eyes or throat or breathing difficulties. Air pollution also contributes to acid rain, smog, haze, and climate change, all of which can drastically affect the environment.

Why should we be concerned about ultraviolet (UV) radiation?

The sun produces three types of UV radiation, much of which is absorbed by the Earth's atmosphere. However, UVA and some UVB are not absorbed and can cause sunburns and other health problems. UV radiation exposure has been linked to health effects including: skin cancers such as melanoma; other skin problems such as premature aging; cataracts and other eye damage; and immune system suppression. Many of these problems, however, can be prevented with proper protection from UV radiation.

Additional EPA resources

- EPA's Office of Air and Radiation: <www.epa.gov/oar/>.

- EPA's Clean Air Markets Web site has information on acid rain: <www.epa.gov/airmarkets/acidrain/index.html>.

- EPA's Office of Transportation and Air Quality has information on air pollution caused by mobile sources: <www.epa.gov/otaq/>.

- EPA's SunWise School Program has information on UV radiation and sun protection: <www.epa.gov/sunwise>.

- EPA's Web site for teachers: <www.epa.gov/teachers>.

- EPA's Air Web site for kids includes information, activities, and games about various issues: <www.epa.gov/kids/air.htm>.

3.2 Water

Why should we be concerned about water quality?

Perhaps the most important problem facing U.S. water bodies today is nonpoint source (NPS) pollution—pollution from many diffuse sources as opposed to one distinct source. NPS pollution is caused by rainfall or snowmelt picking up, carrying, and eventually depositing pollutants into lakes, rivers, wetlands, coastal waters, or underground sources of drinking water. These pollutants include: fertilizers, pesticides, and animal wastes from agricultural lands and residential areas; oil, grease, salts, and toxic chemicals from urban runoff; sediment from improperly managed construction sites, crop and forest lands, and eroding streambanks; minerals from abandoned mines; bacteria and nutrients from livestock, pet wastes, and faulty septic systems; and atmospheric deposition, such as acid rain.

Urban runoff can pose a dual threat to water quality. Natural areas such as forests and wetlands absorb rainwater and snowmelt so that it slowly filters into the ground, reaching waters gradually. In contrast, urban landscapes contain nonporous surfaces like roads, parking lots, and buildings that cause runoff containing toxic oil and grease to increase. Adding to this problem are storm sewer systems that channel large volumes of quickly flowing runoff into a water body, eroding streambanks and damaging streamside vegetation. Native fish and other aquatic life cannot survive in urban streams because of the urban runoff.

Another type of NPS pollution, acid rain deposition, also greatly impacts freshwater environments. When the rate of acids entering lakes and streams is faster than the rate at which the water and surrounding soil can neutralize it, the water becomes acidic. Increased acidity and its associated chemical reactions are highly toxic to many species of fish, insects, plants, and other aquatic species.

NPS pollution has led to beach closures, unsafe drinking water, fish kills, and other severe environmental and human health problems. For example, a large increase of nitrates in drinking water can pose a threat to young children, causing a condition known as "blue baby syndrome." If left untreated, the condition can be fatal. Even adults can be affected by continuous exposure to microbial contaminants at levels over EPA's safety standards. When this occurs, people can become ill, especially if their immune systems are already weak. Examples of the chronic effects of drinking water contaminants are cancer, liver or kidney problems, or reproductive difficulties.

Additional EPA resources
- EPA's Office of Water homepage: <www.epa.gov/OW/index.html>.
- EPA's Office of Water Nonpoint Source Pollution page: <www.epa.gov/owow/nps/>.
- EPA's Office of Water Quality page: <www.epa.gov/ow/national/>.
- EPA's Web site for teachers: <www.epa.gov/teachers>.

3.3 Soil and Land

Why should we be concerned about soil quality?

Soil contamination is a result of either solid or liquid hazardous substances mixing with the naturally occurring soil. Plants can be damaged when they take up contaminants through their roots. Contaminants in the soil can adversely impact the health of animals and humans when they ingest, inhale, or touch contaminated soil, or when they eat plants or animals that have been exposed to contaminated soil. Animals ingest and come into contact with contaminants when they burrow in contaminated soil. Humans can be exposed to toxic elements when they farm, handle, and distribute food and non-food crops. Young children are especially at risk when they play, ingest, or dig in contaminated soil. Certain contaminants, when they contact our skin, are absorbed into our bodies. When contaminants are attached to small surface soil particles they can become airborne as dust and can be inhaled.

Soil contamination can be caused by industrial and chemical byproducts seeping into the soil, spreading metallic substances such as lead, chromium, arsenic, and cadmium. This contamination can also occur from lead-based paints, irrigation, solid waste disposal, fertilizers, and pesticide application. Leaded paint continues to cause most of the severe lead poisoning in children in the United States. It has the highest concentration of lead per unit of weight and is the most widespread of the various sources, being found in approximately 21 million pre-1940 homes. Dust and soil lead—derived from flaking, weathering, and chalking paint—plus airborne lead fallout and waste disposal over the years, are the major sources of potential childhood lead exposure.

Why should we be concerned about land resources?

One of the most pressing land issues in America today is urban sprawl. Sprawl is "the unplanned, uncontrolled spreading of urban development into areas adjoining the edge of a city" (Source: Dictionary.com). This translates to a conversion of rural areas, such as forests and farmlands, into single family homes and strip malls. This type of development uses land inefficiently and increases vehicle miles traveled as people spend more time commuting to and from work.

Another issue affecting American landscapes is that of brownfields and Superfund sites. Superfund is a program administered by EPA to clean up areas where the dumping of chemical and other hazardous wastes might be affecting public health and the environment. Brownfields—abandoned or underutilized industrial or commercial properties with possible environmental contamination—are one type of Superfund site. The cleanup and possible development of brownfields will remove environmental hazards from, and increase the economic well-being of many communities.

Additional EPA resources

■ Information on EPA's Superfund Program:
<www.epa.gov/superfund/index.htm>.

■ Extensive information on brownfields, urban redevelopment news, and resources:
<www.brownfields.com>.

■ The Trust for Public Land, an organization devoted to land conservation: <www.tpl.org>.

■ Information on brownfields on EPA's Web site:
<www.epa.gov/swerosps/bf/index.html>.

4.0 Air-Based Projects

4.1 Teacher Tips

Local air quality affects how we live and breathe. Like the weather, it can change from day to day or even hour to hour. EPA and other organizations make information about outdoor air quality as available to the public as information about the weather. A key tool in this effort is the Air Quality Index (AQI). EPA and local officials use the AQI to provide the public with timely and easy-to-understand information on local air quality. The AQI tells the public how clean or polluted the air is and what associated health concerns they should be aware of. The AQI focuses on health effects that can happen within a few hours or days of breathing polluted air. EPA uses the AQI for five major air pollutants regulated by the Clean Air Act—ground-level ozone, particulate matter, carbon monoxide, sulfur dioxide, and nitrogen dioxide. For each of these pollutants, EPA has established national air quality standards to protect against harmful health effects. The AQI uses a scale of values to indicate the level of health concern and associated color-coded warning. Many EMPACT projects that focus on air quality involve monitoring and collecting near real-time data for the AQI pollutants. In addition, some air projects monitor data related to ultraviolet (UV) radiation, due to its association with stratospheric ozone depletion. For more information on the AQI, go to <www.epa.gov/AIRNow>. For more information on UV radiation and stratospheric ozone depletion, go to <www.epa.gov/ozone>.

Air Quality Index (AQI)*

AQI Number	Health Concern	Color Code
0 to 50	Good	Green
51 to 100	Moderate	Yellow
101 to 150	Unhealthy for sensitive groups	Orange
151 to 200	Unhealthy	Red
201 to 300	Very unhealthy	Purple

*Although ozone reports are primarily made for metropolitan areas, ozone can be carried by the wind to rural areas, where it can cause health problems.

The following are the most common pollutants for which air data is monitored and collected and a description of why the information is important. Throughout this section of the handbook you will read about how this air quality data plays a role in various EMPACT curricula.

- **Ozone (O_3):** Ozone is an odorless, colorless gas composed of three atoms of oxygen. It occurs both in the Earth's upper atmosphere (the stratosphere) and at ground-level. The ozone in the stratosphere is considered "good" ozone because it forms a protective layer that shields us from the sun's harmful UV rays. This ozone is gradually being destroyed by manmade chemicals, such as chlorofluorocarbons. A tool called the UV Index measures the intensity of the sun's rays and can help you plan outdoor activities safely.

  At ground level, ozone is formed when pollutants emitted by cars, power plants, industrial boilers, refineries, chemical plants, and other sources react chemically in the presence of sunlight. Ground-level ozone is unhealthful and is especially problematic during summer months when it is sunny and hot. Ozone can irritate the respiratory system, causing coughing, throat irritation, and/or an uncomfortable sensation in the chest. High risk groups include children or anyone who spends a lot of time outdoors in warm weather and people with respiratory diseases.

- **Particulate matter:** Particulate matter (PM) includes both solid particles and liquid droplets found in the air. Many manmade and natural sources emit PM directly or emit other pollutants that react in the atmosphere to form PM. These particles range in size, with those less than 10 micrometers in diameter posing the greatest health concern because they can be inhaled and accumulate in the respiratory system, causing health problems. Particles less than 2.5 micrometers in diameter are referred to as "fine" particles, and sources include all types of combustion. Particles between 2.5 and 10 micrometers are consider "coarse," and sources include crushing or grinding operations and dust from roads. Coarse particles can aggravate respiratory conditions such as asthma, and exposure to fine particles is associated with several serious health effects, including premature death.

- **Carbon monoxide:** Carbon monoxide (CO) is a colorless, tasteless, odorless gas that forms when the carbon in fuels does not completely burn. The major sources of CO pollution include cars, trucks, and buses; airplanes; trains; gas lawnmowers; snowmobiles; power plants; trash incinerators; and wildfires. CO concentrations are usually highest during cold weather because cold temperatures make combustion less complete and cause inversions that trap pollutants low to the ground. When CO is breathed, it replaces the oxygen that we normally breathe, which deprives the brain and heart of this necessary element. As a result, when exposed to CO, a person might notice shortness of breath or a slight headache. People with cardiovascular disease are most sensitive to risk from CO exposure, and in healthy individuals, exposure to higher levels of CO can affect mental alertness and vision.

- **Sulfur dioxide:** Sulfur dioxide (SO_2) is a colorless, reactive gas that is produced during the burning of sulfur-containing fuels such as coal and oil, during metal smelting, and by other industrial processes. Major sources include power plants and industrial boilers. Children and adults with asthma who are active outdoors are most vulnerable to the health effects of SO_2. The primary response to even a brief period of exposure is a narrowing of the airways, which may cause symptoms such as wheezing, chest tightness, and shortness of breath. When exposure ends, lung function typically returns to normal within an hour. At high levels, SO_2 may cause similar symptoms in non-asthmatics.

- **Nitrogen dioxide:** Nitrogen dioxide (NO_2) is a reddish-brown, highly reactive gas formed when nitric oxide combines with oxygen in the atmosphere. Once it has formed, NO_2 reacts with volatile organic compounds (VOCs), eventually resulting in the formation of ground-level ozone. Major sources of NO_2 include automobiles and power plants. In children and adults with respiratory disease, such as asthma, NO_2 can cause respiratory symptoms such as coughing, wheezing, and shortness of breath. In children, short-term exposure can increase the risk of respiratory illness.

4.2   The Tools

4.2.1   AirBeat (Roxbury, Massachusetts)

Introduction

The AirBeat EMPACT project centers around an air monitoring system—the first of its kind in Massachusetts. The monitoring system, which is sustained by a collaboration of universities, governments, and community organizations, enables residents to check real-time air pollution levels via a telephone hotline or the AirBeat Web site at <www.airbeat.org>. AirBeat measures ground-level ozone and fine particle pollution and focuses on reducing the health effects they have on Roxbury residents, who suffer from high rates of asthma and other respiratory illnesses.

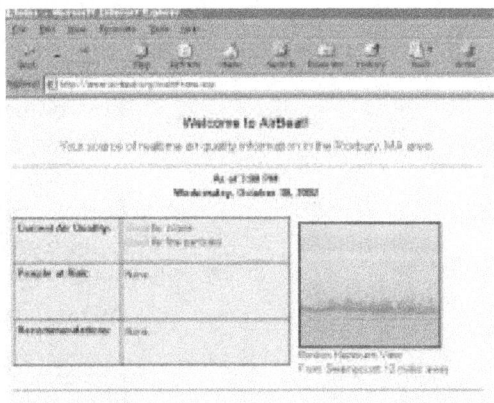

Lessons, Tools, and Activities

Part of the outreach for AirBeat involves educating teachers and students about air quality and its health and environmental effects. Alternatives for Community and Environment (ACE)—a local nonprofit organization—integrated air monitoring into its environmental justice curriculum for local schools by developing an air quality flag warning system that is managed by a local school. Students use AirBeat data to assess air quality on a daily basis and hang flags that correspond to air quality at two locations. The flags advise Roxbury residents about air quality so they can take precautions if they suffer from asthma or other illnesses.

ACE also visits classrooms to administer its air pollution curriculum module, which includes these lessons:

■ *How to Build Your Own Black Carbon Monitor*, adapted from the Lawrence Berkeley National Laboratory, teaches students to build a black carbon monitor from commonly available items and analyze its measurements.

■ Students distribute the *Survey of Air Pollution Awareness* to local residents, then analyze the results to gauge residents' knowledge of air pollution and asthma.

Resources

For more information, contact Jodi Sugerman-Brozan of Alternatives for Community and Environment at 617 442-3343, ext. 23, or <jodi@ace-ej.org> and visit the AirBeat Web site at <www.airbeat.org>, where the above lessons can be downloaded.

Introduction

Air CURRENTS is a curriculum designed to educate middle and high school students about air, air pollution, and air monitoring techniques. The project's name, which stands for Collaboration of Urban, Rural, and Regional Environmental Networks of Teachers and Students, reflects its focus on teachers, students, and learning. The curriculum emphasizes a hands-on, problem-solving approach, after which students implement what they've learned to make changes in the community or region. Teachers and students, in collaboration with community groups, use a portable air monitoring system to do outdoor air monitoring studies in their schools and communities. However, the curriculum can be taught with or without employing the air monitor.

The goal of the Air CURRENTS project is to provide the tools and information necessary for students, teachers, and community-based groups to obtain a general assessment of the air quality in their neighborhoods. Additional goals of the Air CURRENTS program are to integrate environmental learning into core math, science, and social studies curricula; engage students and teachers in scientifically meaningful air monitoring projects; use the Internet to connect participating schools to one another and to resources for air quality and health effects information; and work with schools to aid in developing a community understanding of the complexities of local environmental problems.

The development of the Air CURRENTS curriculum was a collaboration of state and federal agencies, universities, community-based organizations, and educators. The project was managed by Northeast States for Coordinated Air Use Management (NESCAUM), whose purpose is to exchange technical information and to promote cooperation and coordination of technical policy issues among member states. EPA provided a portion of the funding through the EMPACT program to bring the Air CURRENTS curriculum to four EMPACT cities: Buffalo and Brooklyn, NY, and Camden and Newark, NJ.

Lessons, Tools, and Activities

The Air CURRENTS curriculum helps students in grades 6 through 12 understand the causes, consequences, and political complexities of managing air quality. The curriculum is extensive. It contains over 30 consecutive lessons that complete what the Air CURRENTS educators refer to as the full "Science-Technology-Society" (STS) circle. Students complete the STS circle in three steps: (1) gain an understanding of the scientific concepts related to air quality through hands-on laboratory investigations; (2) collect and analyze data after mastering the use of an air quality monitor; and, (3) take appropriate social advocacy actions to support their data and conclusions. Educators believe that since the curriculum actively engages students in a process, it allows them to intimately understand various points of view, so they can create a well-informed opinion about air quality issues for themselves.

The first part of the curriculum introduces important concepts about air—proving that it exists and can be measured, even though students cannot see it.

Students learn about particulate matter and gases such as carbon monoxide. Lessons in the first section provide the conceptual framework for the use of the portable monitor in the second section. Students learn to operate and collect indoor and outdoor air quality data using the ACCESS™ (A Computerized Community-based Environmental Sampling System) portable air quality monitor. After developing a scientific hypothesis and testing it by collecting air quality data using the ACCESS system, students then analyze their data and develop reports describing their findings. While the Web site was active, students posted data files or reports on the Air CURRENTS Web site to share with other students. Students can create a report from a downloaded data file by using the ACCESS™ software from PAX Analytics. Finally, students learn a series of lessons in science, social studies, language arts, math, and arts to complete an advocacy program they could undertake in their community.

Although the curriculum is designed to be used with a portable monitor, the monitor is not required, and segments of the curriculum offer valuable lessons by themselves. The Air CURRENTS curriculum can be taught by a team of teachers across disciplines, but has the flexibility to be taught by science or social studies teachers alone. At the middle school level, the most effective model for this curriculum is where students have designated times for subject areas. At the high school level, teachers have worked in teams of two, either team teaching or working in a parallel model. The environmental sciences are the obvious choices for these curricula, where it can be a self-contained two- to three-month unit, but schools have implemented it into American government, economics, and technology courses.

The Air CURRENTS curriculum utilizes a constructivist approach, which requires teachers to foster an environment for inquiry-based learning. The constructivist approach is based on the premise that human nature dictates that we construct our own understandings of the world in which we live. This approach allows students to actively interact with objects and ideas to test their own preconceptions; then, through reflection of those interactions, develop an understanding. Teachers should establish cooperative learning groups, in which the constructivist model works well. Cooperative learning creates a structured natural environment that promotes collaboration. The teacher, or facilitator in this approach, floats from group to group, to provide guidance as well as ask thought-provoking questions that may encourage their investigations. Students who are exposed to the constructivist model should be given time and space to reflect. Therefore, teachers should encourage students to keep ongoing journals and have an opportunity to reflect on, modify, and redesign their investigations while they are not actively involved in them.

Resources

For more information, or to order a copy of the Air CURRENTS curriculum, contact Susan Green at NESCAUM at 617 367-8540. The NESCAUM Web site <www.nescaum.org> has additional information but does not offer the curriculum for downloading. NESCAUM exchanges technical information and promotes cooperation and coordination of technical policy issues regarding air quality control among member states. They sponsor air quality training programs, participate in national debates on air quality, assist in the exchange of information, and promote research.

The Air CURRENTS Web site <www.aircurrents.org> identifies partners and provides a form for completing the project plan, which can be submitted for review.

Introduction

The Air Info Now project provides current air quality information for the metropolitan Tucson area. The Web site <www.airinfonow.com> was developed under an EMPACT grant along with assistance from the University of Arizona, The American Lung Association, and the Pima Association of Governments. The project site provides information on air pollutants, their health effects, activities to help in understanding air pollution, and historic and current monitoring data.

Tucson, Arizona, is an urban area with a strong public appreciation for and commitment to the surrounding natural environment. The public has shown increasing concern over air pollution, both in terms of individual health and potential environmental impacts in the mountains and high desert lands that are valued locally and worldwide for their pristine condition. Many residents move to the area to alleviate health problems, and therefore, the area has a higher than average percentage of residents who are sensitive to air pollutants. In addition, there are economically disadvantaged areas within the city that have higher documented rates of asthma in children, so the timely dissemination of air pollution data is especially important.

The overall objective of the Air Info Now project is to produce media and public communication programs about air quality, the Tucson environment, health concerns, and local solutions to improve air quality. Other objectives of the project include the following:

- Collecting and disseminating accurate, understandable, and timely air pollution information.

- Expanding associated outreach and education programs to improve understanding of the relationships between air quality, climate, and health effects.

- Allow the community to address local air pollution problems and solutions based on credible scientific information.

The project employs 80 instruments at 18 air monitoring sites throughout the Tucson metropolitan area. In addition to monitoring carbon monoxide, ground-level ozone, sulfur dioxide, nitrogen oxides, and particulate matter (PM10 and PM 2.5), for which EPA has National Ambient Air Quality Standards, the project monitors various meteorological parameters that affect air pollution. These parameters include wind speed, wind direction, temperature, relative humidity, and UV radiation.

Lessons, Tools, and Activities

The Air Info Now project has developed several sets of activities and experiments designed to teach students about pollution prevention, the relationship between air quality and health, and data analysis. The classroom activities offer older students the opportunity to study the health risks that come from ambient airborne pollution in Tucson. The Web site also includes accompanying teacher guides.

Activities (Grades 7 to 12):

Through real-time data collection activities, students learn to analyze and interpret the real-time air quality data that is collected and displayed by the Air Info Now project site. Pollutants investigated include ground-level ozone, carbon monoxide, and particulate matter, and parameters include weather and climate (temperature, wind, rainfall), asthma attacks, visibility, time, and location. Students learn data collection and analysis techniques through practice with Excel spreadsheets and principles of statistics. Students are separated into groups, each representing a different aspect of air pollution. For example, one group represents "location" and tries to identify pollution trends according to location around a city. Another group represents "health effects," and they monitor the occurrences of asthma at several schools to see if there is a correlation with air pollution.

Students regularly share their data with their classmates and summarize their findings in a final paper or project that can be shared with the community.

Experiments (Grades 4 to 12):

Students construct and deploy particulate pollution detectors to test hypotheses: for example, older vehicles and those using leaded or diesel fuel will produce more particulate matter emissions. Students learn to identify gaseous and solid pollutants in the atmosphere; observe an experiment that illustrates how to capture particulate pollutants and identify which vehicle emits more pollutants; and conduct an experiment capturing particulate pollutants and determine which locations appear to have more pollution.

Students make smog in a shoe box or aquarium to demonstrate convection currents and temperature inversion layers and discuss the implications for pollution. They also monitor their family's energy consumption, calculate the amount of carbon dioxide produced, and discover how changes in consumption can affect the amount of pollution and greenhouse gases released.

The Air Info Now Web site also includes several online interactive games for kids that require Macromedia Flash Player.

Resources

For additional information on the Air Info Now project in Pima County or the associated student activities and teacher guides, contact Beth Gorman at Pima County Department of Environmental Quality (PDEQ), 520 740-3343 or <bgorman@deq.co.pima.az.us>. You can download the student activities and experiments, as well as the teacher guides, directly from the Air Info Now Web site at <www.airinfonow.com>. Click on Activities for online games and experiments, and click on Teachers for the data collection activities and teacher guides.

4.2.4 AIRNow (National)

Introduction

Through its Web site, the AIRNow program offers access to daily air quality forecasts as well as real- time air quality data for over 100 cities across the United States. While many EMPACT programs provide the public with easy access to local air quality information, the AIRNow Web site was developed by EPA to offer real-time air quality information for both regional and local areas across the United States and parts of Canada. For example, color maps show ozone levels across a specific regional geographic area. Plus, AIRNow displays air quality forecasts (good, moderate, unhealthy for sensitive groups, unhealthy) for "air action days" in major metropolitan areas around the country. Users can view local or regional air quality information such as ozone maps and air quality forecasts and learn more about how they should adjust their outdoor activity level when air quality is forecast to be poor. The Web site links to more detailed state and local air quality Web sites.

A central component to the daily air quality forecast is the Air Quality Index, or AQI. (See Section 3.1 for more on AQI.) The AIRNow Web site uses the AQI categories, colors, and descriptors to communicate information about air quality. Increasingly, TV, radio, and newsprint forecasters are providing information using the AQI. During summer months, for example, you may learn that it is a code red day for ozone, meaning the air quality is unhealthy. But how do you know what this means? Parents can learn by visiting the AIRNow Web site and reading about the AQI. To help teach children how to read and understand the AQI, the Web site offers an online and downloadable curriculum for school-aged children.

Lessons, Tools, and Activities

The AIRNow curriculum is geared toward children 7 to 10 years old. EPA developed more ozone segments for the 2002 ozone season (May through October), aimed at those 5 to 6 years old, as well as those 7 to 10 years old. A Spanish version of the current curriculum was launched in March 2002.

The AIRNow lessons can be used online or teachers can print a text version of the Air Quality Index Kid's Web site and curriculum for classroom use. The kids page includes two animated online games that can also be printed. The animated version requires a Flash 5 plug-in player, which is available on the Web site.

Lessons (Grades 2 to 5):

The Kids section of the AIRNow Web site is hosted by an animated trio of chameleons: K.C. Chameleon, Koko Chameleon, and Kool Chameleon. Kids navigate through four topic areas, learning about the AQI, clean and dirty air, and how health is affected by breathing dirty air. By viewing an animated cartoon, kids learn that ozone is formed by a combination of pollution and sunlight. They also learn where soot and dust come from and how particulate matter is formed. Once they learn about pollutants and how they affect our bodies, they learn how EPA and local governments present this information to the public using the AQI.

By navigating different parts of the AIRNow Web site, kids find the AQI forecast and an ozone map for their area. They learn the numbers, colors, and words that the AQI uses to describe air quality. By learning to identify groups that are sensitive to ozone—asthmatics, children, and the elderly—they can read an AQI forecast and understand what those groups should do differently on poor air quality days. Finally, kids learn what they can do to reduce pollution and improve air quality.

As kids navigate, they have the opportunity to explore and further their learning. As they encounter new words, each page links to a dictionary of air pollution related words such as "global", "pollution", and "smog". They also learn where on the Web site they can view ozone maps covering their local area. The Web site includes two games: AQI Color Game and the AQI Game Show. The AQI color game contains three levels of difficulty, from the easier word and color connecting game, to the more challenging game, in which an AQI numerical value is given and kids must look up the corresponding color.

In the AQI Game Show, three chameleons play the contestants, answering multiple choice questions about AQI and health. Kids click on the chameleon with the correct answer, and the game automatically keeps score. The online version includes 10 questions and the printed version includes 27 questions. The answers are provided and both games can be downloaded and played on hard copies.

From the AIRNow Web site, teachers can print colorful posters for each of the five most common color codes of the AQI. For each color code, one of the chameleons tells kids what level of outdoor activity is recommended for them that day. The posters will print in color on a color printer. For schools without color printers, a good exercise could be to color the posters the correct color. Teachers can contact the AIRNow program to request color copies.

For more information on AIRNow, contact John E. White of EPA at 919 541-2306 or at <white.johne@epamail.epa.gov>. The entire curriculum can be downloaded from the AIRNow Web site at <www.epa.gov/airnow/aqikids/teachers.html>.

4.2.5 Community Accessible Air Quality Monitoring Assessment (Northeast Ohio)

Introduction

The Northeast Ohio (NEO) EMPACT project focuses on developing an improved air monitoring data and network system and creating a land-use data and ecological computer modeling tool. The latest technology provides communities with real-time air quality reports. These help citizens make informed decisions on everyday quality of life issues such as environmental and health concerns that accompany urban living and growth. The NEO EMPACT air quality project also conducts community outreach to inform citizens about NEO air quality programs and resources.

Lessons, Tools, and Activities

As part of the NEO EMPACT project, a handbook was developed to introduce teachers and students to the importance of understanding air quality in their communities. Air Quality in Northeast Ohio is arranged in thematic and developmental order to provide students with a comprehensive understanding of air quality and its effects on health and the environment.

The 85-page handbook for educators and 4th through 8th grade students includes detailed background information, lessons, and activities focused on air quality. It progresses from conceptually developing an understanding of air quality to discussing concrete actions students and teachers can take to improve air quality. The main sections of the handbook include:

- **Educator's Notes** includes background information that prepares educators to administer the lessons and exercises in the handbook. The section describes air pollution, its origins, and its health and environmental effects. It also contains information on the Clean Air Act, specific pollutants (e.g., carbon monoxide, particulate matter), acid rain, and the effects that vehicles and weather have on air quality.

- The 10 **Experiments and Exercises** give students hands-on lessons in air quality. Geared towards specific grades, the exercises and experiments cover air quality vocabulary, visible and invisible air pollutants, smog, air pollution's effects on plants, and air quality data analysis and tracking.

- The **Internet-based Activities** teach students to access NEO EMPACT air quality data online.

- The **Air Quality Activities** focus on developing students' oral, visual, and writing skills. Activities include conducting a mock interview with an environmental professional, writing a clean air bill, composing a letter about acid rain for local legislators, completing air quality word searches and crossword puzzles, and designing air quality posters for display in the community.

- **Reducing Air Pollution—What Students Can Do** offers teachers and students some suggestions for reducing air pollution in the local community and at home.

- **Air Quality Resources and Materials for Educators** lists additional Internet, hard copy, and organizational resources for air quality information. It also includes ideas for no-cost educational materials and how to obtain them.

Resources

To obtain a free copy of the NEO Air Quality Curriculum Handbook, contact Adam Zeller of the Earth Day Coalition at 216 281-6468 or <azeller@ earthdaycoalition.org>. For more information on the NEO EMPACT project, visit the NEO EMPACT Web site at <http://empact.nhlink.net> or the Northeast Ohio Air Quality Online Web site at <http://neoair.noaca.ohiou.edu>.

4.2.6 ECOPLEX (Dallas-Ft. Worth, Texas)

Introduction

Through the use of both innovative and proven environmental monitoring technologies, the ECOPLEX project collects real-time and time-relevant environmental data that informs citizens of the Dallas-Ft. Worth metropolitan area of current, historical, and near real-time forecasts of environmental conditions. The project involves a multimedia approach, collecting data related to air, water, soil, and weather. The data, as well as instructions on how to use it, are posted on the project's Web site at <www.ecoplex.unt.edu>.

Lessons, Tools, and Activities

As part of the ECOPLEX project, curricula were developed covering the topics of ultraviolet (UV) radiation, water quality, and water quantity. (See Section 5.0 Water-Based Projects for information on ECOPLEX water lessons.) The curricula are geared towards kindergarten through 8th grade and were completed in August 2001. Approximately 120 teachers in 37 schools have utilized the lesson plans included in the curricula.

Each lesson plan includes follow-on curriculum extensions, which explore the disciplines of math, language arts, technology, art and music, science, and social studies.

The air portion of the ECOPLEX curriculum introduces students to the dangers of UV rays and the connection to stratospheric ozone. Through simple, yet progressively challenging experiments, lessons, and activities, children in grades kindergarten through 3 learn ways to protect themselves from harmful UV rays and to develop a daily routine of UV protection, similar to brushing their teeth. Students learn about the shadow rule—if your shadow is taller than you, UV exposure is usually low, and if it is shorter than you, UV exposure is usually high—and ways to identify sun-safe areas on the playground. They are introduced to the ECOPLEX Web site and learn how to read the UV Index. Children witness how UV rays are affected by the time of day and the seasons, and they learn to identify the layers of the atmosphere, discussing how stratospheric ozone is depleted. They develop plans for reducing their personal exposure to UV rays and set goals for how they can reduce the formation of ground-level ozone.

Students in grades 4 through 6 learn that stratospheric ozone blocks UV rays and that certain materials deplete this type of ozone. Using the UV meter, students determine the dangers due to UVA and UVB and measure UV levels throughout the day. Then they create a comparison between the UV meter readings and ECOPLEX UV data over a period of time, graphing the results. Students explore the electromagnetic spectrum, finding where UV light fits in, and they view the refraction of light using a prism, identifying the invisible rays: infrared, heat waves, and UV rays. Using bacteria culture, students observe which types of light best prevent bacteria growth. With their findings, students create an informative brochure to distribute to family and friends.

In grades 7 through 8, the ECOPLEX curriculum helps students understand how the angle of the sun on earth affects temperature. They conduct light experiments using a flashlight on a world map to mimic the sun on the earth, and they record their estimations of direct and indirect solar energy, demonstrating how direct solar energy is affected by the seasons and the time of day. Children learn about how chlorofluorocarbons (CFCs) destroy ozone through chemistry experiments and they become aware of how the use of certain products releases CFCs into the atmosphere.

Resources

For more information on the ECOPLEX UV curriculum, contact Ruthanne (Rudi) Thompson at <rudi@unt.edu> or 940 565-2994 and visit the ECOPLEX Web site at <www.ecoplex.unt.edu>. Click on the Teacher's Corner to download lessons as PDF files.

4.2.7 SunWise School Program (Nationwide)

Introduction

The SunWise School Program is a national environmental and health education program that aims to teach children in grades kindergarten through 8 and their caregivers how to protect themselves from overexposure to the sun. Through the use of classroom-based, school-based, and community-based components,

SunWise seeks to develop sustained sun-safe behaviors in schoolchildren and foster an appreciation of the environment around them.

The program's leading components build on a solid combination of traditional and innovative education practices already in use in many U.S. elementary and middle schools. Through the program, students and teachers increase their awareness of the harmful effects of ultraviolet (UV) radiation and learn simple ways to protect themselves and their family. Children will also acquire scientific knowledge and develop an understanding of the environmental concepts related to sun protection.

The program encourages schools to implement a sun-safe infrastructure, including shade structures, such as canopies and trees, and policies, such as using hats, sunscreen, and sunglasses on a regular basis. Designed to provide maximum flexibility, the SunWise program elements can be used as stand-alone teaching tools or to complement existing school curricula. Registering to become a SunWise school can easily be accomplished on the SunWise Web site at <www.epa.gov/sunwise>.

Lessons, Tools, and Activities

A useful resource for SunWise school partners is the SunWise Tool Kit, which contains cross-curricular lessons and background information for kindergarten through 8th grades. The Tool Kit consists of a variety of fun, developmentally appropriate activities that combine education about sun protection and the environment with other aspects of learning. The SunWise Web site, a very helpful tool, provides downloadable information, storybooks, and activity books, some of which are available in Spanish. The SunWise curriculum includes age-appropriate, progressively challenging material to teach students of all levels the importance of sun protection.

Younger students in kindergarten through 2nd grade are introduced to the concept of UV rays and their potentially harmful effects, and they begin to learn simple ways to protect themselves from the sun. They make wacky sunglasses out of paper and cellophane in various colors to emphasize the importance of wearing sunglasses. Educators tell fun stories and legends about the sun and play interactive games like "Sunny Says," following the format of "Simon Says." Students learn which products at the store are sun safe, and they participate in activities such as shadow tracing, which introduces the importance of the "No shadow, seek shade" rule. Using maps, magazines, and photos of various places and peoples around the world, children learn that numerous societies practice sun safety in a variety of ways.

Intermediate students in 3rd through 5th grades perform word games such as word scrambles and crossword puzzles using keywords that emphasize sun safety and protection. The SunWise Tool Kit provides a special UV sensitive frisbee that changes color when exposed to UV radiation. As an experiment, students place different materials, such as tanning lotion and sunscreen, onto the frisbee and expose it to the sun. The students watch as the unprotected portions of the frisbee change color and the protected areas remain the same; they then record their findings on a data chart. Students have the opportunity to go on the

Internet and discover the variety of existing sun myths, understanding how different cultures perceive the origins and history of the sun. They learn the difference between "good" and "bad" ozone, and perform experiments such as witnessing the sun's effects on fruit and newspapers. They assess the risk factors of their own skin and put on a SunWise fashion show, identifying the differences between sun safe and unsafe clothes.

Students in grades 6 through 8 perform numerous activities that correspond to a variety of subjects. They brainstorm, using their creativity and imagination to write songs, public service announcements, and news stories exploring the risks of UV exposure. They create a puppet show to teach younger school kids about protecting themselves from the sun. They act as architects and submit a design proposal for a new SunWise playground. Through Internet searches, students deepen their understanding of the various cultures and myths around the world, going on virtual vacations, picking destinations and identifying sun safe items to pack in their suitcases. They research skin cancer statistics and interpret their findings state by state. They pretend they are Galileo or Copernicus and write journal entries about their beliefs and what the future will be like. Seasonal Affective Disorder (SAD), the disorder applied to people who suffer depression during winter, is explored and discussed, and students reexamine the benefits and the risks of sun exposure.

Resources

For additional information on the SunWise School Program, visit <www.epa.gov/sunwise> or contact Kristin Kenausis of EPA at 202 564-2289. Only K-8 schools who register for the program can receive the Tool Kit, but many other educational materials and publications are available for downloading from the Web site or from the clearinghouse (800 490-9198). Visit the "Publications" page on the SunWise Web site for more details.

5.0 Water-Based Projects

5.1 Teacher Tips

Scientists that study lakes and reservoirs—limnologists—are interested in obtaining data for several water quality parameters. Many of these parameters can be measured remotely, without having to bring samples to a laboratory for analysis. The following are the most common parameters for which data is collected and a description of why the information is important. Throughout this section of the handbook you will read about how this water quality data is utilized in various EMPACT curricula.

- **Chlorophyll:** Chlorophyll are complex molecules found in all photosynthetic plants, including aquatic plants called phytoplankton. Chlorophyll allows plants to use sunlight as part of their metabolism. The distribution and concentration of phytoplankton is of major water quality and ecologic concern. Certain inputs of critical plant nutrients, such as phosphorus, can lead to excess concentrations of phytoplankton. Because the amount of phytoplankton affects the clarity and color of water in lakes and reservoirs, it is of concern to scientists and environmental managers. The most common method of determining the amount of phytoplankton in a body of water is to measure chlorophyll concentration, which is done either by using an analytical/instrumentation technique (e.g., spectrophotometer, fluorometer, high-pressure liquid chromatography) on filtered samples or using fluorescence technology, which allows for semi-quantitative measurement of chlorophyll in phytoplankton cells without extraction or chemical treatment, thereby allowing in situ (in-lake) measurements.

- **Turbidity:** Turbidity refers to the extent that water lacks clarity. It is therefore, tightly linked with the aesthetics and perception of water because the public wants water of high clarity for recreation. Turbidity is caused by a mixed population of suspended particles, which may include clay, silt, finely divided organic matter (detritus), phytoplankton, and other microscopic organisms. In general, these particles are a composite of sediments received from inflowing tributaries, resuspended sediments, and particles produced within the body of water (particularly phytoplankton). Thus, the variations in measured turbidity may reflect the dynamics of phytoplankton growth as well as tributary runoff (driven by rainfall events). Until recently, turbidity was measured using a nephelometer, where a beam of light is directed along the axis of a cylindrical glass cell containing the sample. Light scattered by particles from the beam is measured by a detector. New technology has led to the development of turbidity probes that can be constructed on remote sampling units. These probes are constructed in a similar manner as the nephelometer, except that the scattered light detector is located within the water as opposed to outside a glass sample cell.

- **Temperature:** Temperature is a measure of molecular vibrational energy. It has extremely important ecological consequences. Temperature exerts influence on aquatic organisms with respect to selection and occurrence and level of activity of the organism. In general, increasing water temperature results in greater biological activity and more rapid growth. All aquatic organisms have a preferred temperature in which they can survive and reproduce optimally. Temperature is also an important influence on water chemistry, as rates of chemical reaction increase with increasing temperature. Temperature regulates the solubility of gases and minerals (solids)—warm water contains less dissolved oxygen and more solids than cold water. Thermal stratification refers to the layering that occurs, particularly in the warm months. Typically, a warmer, less dense layer called the epilimnion overlies a colder, denser layer called the hypolimnion. In between these two layers is a third layer called the metalimnion where strong differences in temperature and density exist. Seasonal changes cause mixing of the layers. Usually, a thermometer is used to determine temperature, although when taking measurement below the surface, methods such as thermocouples and thermistors can be used. A thermocouple measures the current generated by two different metals at different temperatures. A thermistor measures voltage produced by a semi-conducting material that decreases in resistance with increasing temperature.

- **Conductivity:** Electrical conductivity is a measure of water's ability to conduct electricity, and is therefore a measure of the water's ionic activity and content. The higher the concentration of ionic (dissolved) constituents, the higher the conductivity. Wide variations in water temperatures affect conductivity, making it difficult to make comparisons of this feature across different waters, or changes in this parameter for a particular body of water. The use of specific conductance, which is the conductivity normalized to 25° C, eliminates this problem and allows comparisons to be made. Specific conductance is a reliable measure of the concentration of total dissolved solids (TDS) and salinity. It also is a valuable tracer of water movement. By definition, specific conductivity is the reciprocal of the specific resistance of a solution measured between two electrodes (opposite electrical charges) placed in the water. For a known electrical current, the voltage drop across the electrodes reveals the water's resistance. Since the resistance of aqueous solution changes with temperature (resistance drops with increasing temperature), the resistance is corrected to the resistance of the solution at 25ºC.

- **Dissolved Oxygen:** The concentration of dissolved oxygen (DO) is probably the single most important feature of water quality, as it is an important regulator of chemical processes and biological activity. Plant photosynthesis produces oxygen within the region below the water surface with adequate light. Microbial respiration and organic decay consume oxygen. At the surface, oxygen can move between the water and air, and the rate of exchange is dependent on wind speed and the surface water DO saturation. The saturation concentration of DO is regulated by temperature. Concentrations above the saturation value (supersaturation) indicate high photosynthetic activity, for example, during an algal bloom. Undersaturated conditions occur when oxygen-demanding processes exceed the sources of DO. DO is measured using a probe that consists of electrodes of opposing charges, which are separated from the surrounding water by a Teflon membrane. DO diffuses across the membrane and is reduced to hydroxide at the cathode and silver chloride is formed at the anode. The current associated with this process is proportional to the DO in the surrounding water.

- **pH:** pH is defined as - log [H+], where [H+] = concentration of hydrogen ions. The pH scale ranges from 0 to 14, corresponding to various degrees of acidity or alkalinity. A value of 7 is neutral; values below 7 and approaching 0 indicate increasing acidity (higher H+ concentrations), while values above 7 approaching 14 indicate increasing alkalinity. A wide range of pH values is encountered in different water bodies, associated primarily with the different ionic chemistries of the respective watersheds/tributaries. Inorganic carbon constituents are the major pH buffering system in most fresh waters. pH is an important regulator of chemical reactions and an important influence on aquatic biota (including composition). Photosynthetic uptake of CO_2 tends to increase pH (e.g., during phytoplankton blooms) while decomposition/respiration tends to decrease pH. Values of pH are generally highest in the epilimnion and decline with increasing depth. Measuring pH involves taking an electrode consisting of a proton selective glass reservoir filled with a pH 7 reference solution. Protons interact with the glass, setting up a voltage potential across the glass. Since the H+ concentration of the reference solution does change, the difference between the voltage potentials is proportional to the observed pH.

5.2   The Tools

5.2.1   Boulder Area Sustainability Information Network (BASIN) (Boulder, Colorado)

Introduction

The Boulder Area Sustainability Information Network (BASIN) project is an EMPACT-funded project designed to help deliver a variety of environmental information about the Boulder area to its residents. BASIN's initial focus is on water in the region, including watershed and consumption issues. The objectives of the project include the following:

- To improve existing environmental monitoring to provide credible, timely, and usable information about the Boulder Creek Watershed to the public.

- To create a state-of-the-art information management and public access infrastructure using advanced, Web-based computer technologies.

- To build strong partnerships and an ongoing alliance of governmental, educational, nonprofit, and private entities involved in watershed monitoring, management, and education.

- To develop education and communication programs to effectively utilize watershed information in the public media and schools and facilitate greater public involvement in public policy formation.

Lessons, Tools, and Activities

As part of the project, organizers adapted an existing online learning tool called the WatershED program, to the BASIN project Web site. Geared toward grades 4 through 12, WatershED aims to help teachers, students, and citizens in the Boulder area learn more about their local creeks and wetlands. It provides users with suggestions for what schools or neighborhood groups can do to preserve and protect local waterways and how they can become stewards of water resources.

The WatershED curriculum was developed by the Boulder Creek Initiative and the City of Boulder's Stormwater Quality Office with the help of teachers in the Boulder area. It was modified for students, teachers, and the general public for the BASIN Web site. The tool consists of a series of learning activities in addition to a Teacher's Guide.

The WatershED project can help participants:

- Get to know their watershed address as defined by creeks, wetlands, and lakes.

- Discover the plants, animals, and birds they might see in or around the creek or wetland in their neighborhood.

- Organize a StreamTeam to protect and enhance a local waterway.

The online resource includes background information on ecology and ecosystems and water quality. The activities cover the following topics, which are broken out by level of complexity as follows:

Introductory Level Activities:

- Water, Colorado's Precious Resource

- The Water Cycle

- The Boulder Water Story

- Water Law and Supply

- Water Conservation

Intermediate Level Activities:

- Stream Teams—An Introduction

- Mapping Your Watershed

- Watershed Walk

- Watershed Cleanup: A Treasure Hunt

- Storm Drain Stenciling

- Raise and Release: Aquarium Setup

Advanced Level Activities:

- Water Quality (Introduction)

- Phytoplankton—Trends & Diversity

- Nutrients: Building Ecosystems in a Bottle

- Macroinvertebrates—Long-term Ecosystem Health

- Stream Gauging: A Study of Flow

- Water Quality (Intermediate and Advanced)

Resources

For additional information on the WatershED online learning tool affiliated with the BASIN EMPACT Project, contact Curry Rosato at 303 413-7365 or Donna Scott at 303 413-7364. In addition, all the activities listed above are available online at <bcn.boulder.co.us/basin/learning/introduction.html>.

5.2.2   Burlington Eco Info (Burlington, Vermont)

The goal of the Burlington Eco Info EMPACT project is to provide the public with clearly communicated, real-time, useful, accurate environmental monitoring data in an ongoing and sustainable manner. The project is a 2-year pilot project that will enable residents and policymakers alike to have expanded access to important environmental information, providing for improved decision-making. The project's partners include the City of Burlington Community and Economic Development Office, the University of Vermont (UVM) School of Natural Resources, the Green Mountain Institute for Environmental Democracy, the Center for Lake Champlain (formerly called the Lake Champlain Basin Science Center), and the U.S. Environmental Protection Agency. The project's Web site provides information on the air, water, land, and energy in Burlington and the surrounding area. Visitors can learn about city beaches, view the daily air quality forecast, see a live image of the waterfront, or get data from a dust monitoring station.

Lessons, Tools, and Activities

Although the Burlington Eco Info project is multi-media in nature, the curriculum portion of the project focuses on water quality issues in the Lake Champlain Basin. Through its partnership with the Center for Lake Champlain, the project has incorporated an environmental monitoring program for grades 7 through 12. The program utilizes the UVM's Ecosystem Science Lab (Rubenstein Lab) to perform analyses. The purposes of the environmental monitoring program are the following:

- For students and teachers to participate in and perform authentic scientific research techniques in a university lab setting.

- To promote watershed awareness and action focusing on water quality issues in the Lake Champlain Basin.

- To collect data and allow teachers and students to become involved with local watershed resources with the goal of contributing data that meets EPA's standards for water quality testing.

- To build stronger connections between students and teachers and their local watersheds.

The Center for Lake Champlain markets the program to middle and high school science educators in Vermont and New York schools and organizations located in the Lake Champlain watershed. Interested educators sign up for a teacher training led by the Center staff. After the training, teachers begin the program by teaching water quality related scientific activities at their schools. Following these activities, the class collects local water samples and visits the Rubenstein Ecosystems Lab at UVM to process them. Teachers and students then return to their schools for completion of the processing of their data and other followup activities.

Pre-visit Activities at School

Prior to taking the water samples, students use activities provided by the program and/or found in existing curricula (This Lake Alive!, Project Wild, Project WET, Aquatic WILD, etc.) to get necessary lab skills and knowledge of ecological principles. Trained UVM Resource Assistants visit classrooms to go over safety procedures and understanding of watershed issues. An interactive watershed model is used to help students visualize watershed concepts. In addition, students explore the geography of the Lake Champlain Basin and study the properties of water (pH, water cycle, etc.) to build a stronger connection between the field, lab work, and environmental health. Finally, students generate a focus question for their study.

Field Work Component

Students and teachers collect water samples and other information from a local site of their choice according to established protocols. Teachers also have the option to add a waterfront field component to their class time spent with the Center staff. The 1-hour waterfront option explores "in the field" sampling techniques and includes parameters such as temperature, pH, dissolved oxygen, conductivity, and turbidity.

Rubenstein Lab Activities

In the lab, students perform high-level tests on the water samples they collect in the field. The data generated by the tests are sent to local, state, and federal databases. The first part of the lab activity begins with students practicing lab techniques using glass and plastic pipettes and droppers. Through simple activities, such as color mixing and water drops on a penny, students immediately become actively engaged in the learning process. More sophisticated water sample analysis follows, which includes phosphorous and bacteria testing and a slide presentation designed for the program.

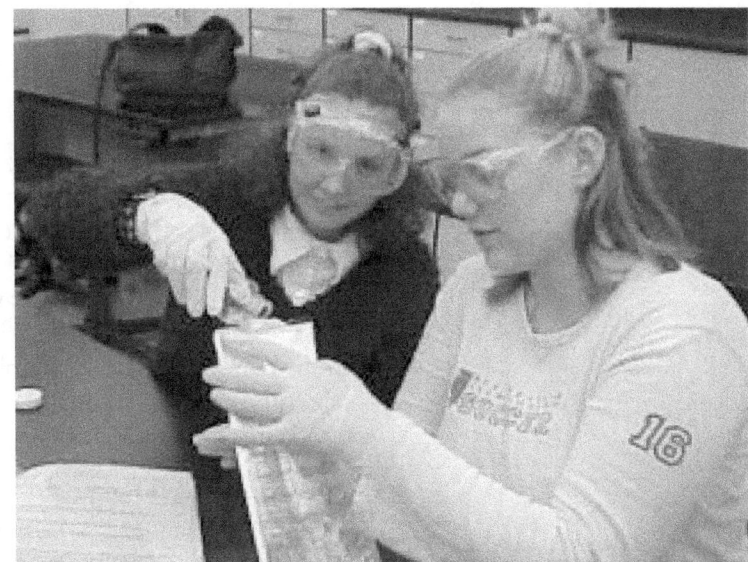

To date, 28 educators from 16 different schools and organizations in the Champlain Basin have participated in the teacher workshop in preparation for bringing their classes to the Rubenstein Lab to conduct water testing, and 267 middle and high school students from 13 schools have participated in the environmental monitoring program. Through three postcard and flyer mailings sent during fall 2000, fall 2001, and spring 2002, the Center reached more than 750 Vermont and New York middle and high school educators.

The Center for Lake Champlain offers a Watershed Investigation Kit for interested teachers, which was not funded through EMPACT, but rather a different EPA grant. The Kit contains everything needed for a thorough water quality study, including books, articles, maps, posters, videos and CD-ROMs, flashcards, and sampling test kits and materials. The Kit is recommended for middle and high school students and for community groups to use in asking questions and discovering more about their place in the Lake Champlain watershed.

Resources

For additional information on the environmental monitoring curriculum offered by the Center for Lake Champlain in association with the Burlington Eco Info EMPACT project, contact Julie Silverman at 802 864-1848 or <juliesilverman@yahoo.com>, or Kara Lenorovitz at <klenorovitz@hotmail.com>.

5.2.3   ECOPLEX (Dallas-Ft. Worth, Texas)

Introduction

Through the use of both innovative and proven environmental monitoring technologies, the ECOPLEX project collects real-time and time-relevant environmental data that informs citizens of the Dallas-Ft. Worth metropolitan area of current, historical, and near real-time forecasts of environmental conditions. The project involves a multi-media approach, collecting data related to air, water, soil, and weather. The data, as well as instructions on how to use it, are posted on the project's Web site at <www.ecoplex.unt.edu>.

Lessons, Tools, and Activities

As part of the ECOPLEX project, curricula were developed covering the topics of ultraviolet (UV) radiation, water quality, and water quantity. (See Section 4.2.6 for information on the UV curriculum.) The curricula are geared towards kindergarten through 8th grade and were completed in August 2001. Approximately 120 teachers in 37 schools have utilized the lesson plans included in the curricula. Each lesson plan includes follow-on curriculum extensions, which explore the disciplines of math, language arts, technology, art and music, science, and social studies.

For kindergartners through 3rd grade, the ECOPLEX curriculum teaches students the quality, importance, and availability of water to life on earth. Students are introduced to the term "water quality" and learn the difference between drinking, fresh, and salt water. They learn how much of people's bodies and certain foods, such as fruit, consist of water. While visiting the ECOPLEX Web site to study water monitoring tests, students brainstorm ways to create good water quality. Students explore the dehydration process in foods, and they learn about precipitation, evaporation, and condensation, and how water can be a solid, liquid, or gas. Introduced to the concept of water conservation, children realize that the amount of water on earth is finite and that most of it is not available for public consumption. They discover how all the water we use is piped to a wastewater treatment system, so that it can be reused. They learn the differences between point and nonpoint source pollution and the physical and chemical aspects of water. And finally, they study the formation of reservoirs and lakes and discover the importance of wetlands as natural filters.

Intermediate students in 4th through 6th grades are introduced to the concepts of food webs and chains. Students learn how pollutants can enter water, affect aquatic organisms, and disrupt food chains. The curriculum covers topics such as groundwater and aquifer recharge, allowing students to discuss from where they get their water and chemical pollutants that cause serious concern, such as DDT, polychlorinated biphenyls (PCBs), and mercury. They discuss bioaccumulation and describe how DDT entered the eagle food chain. Water conservation is reemphasized, as students discuss ways that families can conserve. Students learn how aquatic organisms get oxygen, define photosynthesis and its reliance upon sunlight, and determine the effect of temperature on dissolved oxygen.

Older students in 7th and 8th grade further examine water quality by analyzing macroinvertebrates in the water. They learn that an ecosystem is a community of living and non-living components and that photosynthesis is important to both plants and animals. Students then conduct an experiment to see how fertilizers affect algae growth in bodies of water. Through collecting water samples from a local source, students record the numbers of macroinvertebrates and determine water quality. The ECOPLEX

WOW is highlighted in this handbook because of its affiliation with the Lake Access EMPACT project <www.nrri.umn.edu/empact>—in 2000, EMPACT funded the deployment of two additional RUSS units in Lake Minnetonka, a large, heavily used complex in the suburban Minneapolis area.

curriculum enables students to determine where their water comes from and the quantity of water used by individuals, families, and cities. Students learn about alternative solutions for future fresh water supplies, building upon previous lessons on the watercycle, watersheds, surface water, and fresh water conservation. Using world maps or globes, students discuss how water is redistributed around the globe via the watercycle, and they discuss the effects of population on water supplies and alternative solutions to collect and store water.

5.2.4   Lake Access (Water on the Web) (Minneapolis, Minnesota)

Introduction

Water on the Web (WOW) is a National Science Foundation-funded, award-winning, Internet-based science curriculum for high school and college level students. The project, operated by the University of Minnesota-Duluth's Natural Resources Research Institute, uses real-time, environmental lake data with the goal of equipping students with real world skills they can use in college and beyond. The program employs several remote underwater sampling stations, or RUSS units, in four Minnesota lakes and bays that represent a wide range in terms of size, depth, seasonal dynamics, and other characteristics. The RUSS units collect vertical profiles of temperature, dissolved oxygen, pH, conductivity, and turbidity every few hours and upload their data onto the WOW and Lake Access Web sites each morning.

WOW is based on real, scientific data, monitored and maintained by quality control protocols. Unlike canned data sets created to support a curriculum, the WOW data reflect the realities and complexities of real ecosystems, which means they do not often fit students' or teachers' preconceived ideas of how a lake behaves. WOW data are provided in several different formats in the data section of the WOW Web site. Raw data for a lake can be viewed in an archived data set. Weekly data sets can also be downloaded and reviewed in Excel spreadsheets, which also include graphing templates that assist students in plotting and understanding selected data. For many students, however, it is difficult to see and interpret patterns in numerical data, so WOW offers interactive data visualization tools. Some teachers use these tools to illustrate trends or relationships among the data, and other teachers have students explore the data using the tools. To provide students with the background information and context for understanding scientific data, the WOW Web site includes a variety of aids, including the following:

- Background information on each lake, its watershed, and its behavior during the period of sampling.

- A Lake Ecology Primer, which provides a context for understanding water quality parameters and how they relate to each other.

- A Geographic Information Systems (GIS) resource that describes the fundamentals of the technology.

- A section called "The RUSS," which provides students with an introduction to RUSS technology, WOW water quality measurements, reporting limits, and instrument accuracy.

- A glossary providing definitions of complex scientific terms.

Lessons, Tools, and Activities

The WOW curriculum provides a collection of individual, yet integrated, lessons designed to enrich and enhance student learning in general science courses. Most lessons appear in two different formats—a "Studying" lesson and an "Investigating" lesson. "Studying" lessons allow students to apply and learn concepts through direct, guided experiences. "Investigating" lessons provide students with opportunities to discover the same concepts and involve more solving. Each lesson is organized into a thinking framework of six sequential parts that are critical for improving scientific and technological literacy—knowledge base, experimental design, data collection, data management and analysis, interpretation of results, and reporting results. Using this format for scientific inquiry, teachers guide students through directed study or inquiry lessons depending on the students' abilities and the science curriculum.

Messages from teachers indicate the WOW lessons and Web site are being used in a variety of ways. One teacher used a tutorial and lessons to help students learn how to work with spreadsheets. Another adapted a lesson on fish stocking to illustrate that organisms are limited by environmental factors. Still other teachers have chosen ideas from the lessons and Web site and created their own lessons based on WOW data and resources.

"I found the Water on the Web site to be of great value and interest to the students...It was a wonderful source of detailed information and provided the students with access to nearly real-time water quality data. I was able to use the information to devise very realistic problems for the students to work through and discuss."

—George W. Kipphut,
Murray State University, Kentucky

"Thank you for the wonderful data and project...This project puts symmetry on the year for us...The focus and quiet as they delve into the data and resources are great."

—Ilona Rouda,
The Blake School,
Minneapolis, Minnesota

Since the program's inception in 1998, several thousand students have used WOW and its materials. Students have learned the fundamentals of science based on real-time data, and teachers have been trained in advanced technology, including computerized mapping and modeling systems, remote sensing, instrumentation, and the use of the Internet.

A project is currently underway to create an online curriculum geared toward college students in 2- to 4-year institutions. This curriculum will serve as a capstone experience for students who are completing a technician program, or a gateway for students who are stimulated by the issues and interested in pursuing water science, water resource management, or environmental resource management degrees at four-year institutions. Students will learn and apply their knowledge and skills through inquiry-based problems derived from real-world, real-time data collected by state-of-the-art water quality monitoring technology. The curriculum will be designed as a two semester lab sequence, consisting of six key units that cover the range of knowledge and skills needed by future water science technicians. Each unit will consist of a series of 3 to 8 interactive modules that cover specific topics (e.g., the Data Analysis Unit will include Web-based modules on Exploratory Data Analysis, Trend Analysis, Spatial Analysis, and Modeling). The curriculum will receive extensive pilot and beta testing by a group of over 100 community college teachers and will be designed to be disseminated through a commercial publisher.

Resources

For more information on the WOW project and curricula, contact:

George E. Host, Ph.D.
Senior Research Associate
Biostatistics-Forest Ecology
University of Minnesota-Duluth Campus
Center for Water and the Environment
Natural Resources Research Institute
Phone: 218 720-4264
Fax: 218 720-4328
E-mail: <ghost@nrri.umn.edu>

or

Bruce Munson
University of Minnesota-Duluth Campus
Phone: 218 726-6324
E-mail: <bmunson@d.umn.edu>

WOW information and lessons are all downloadable from the project's Web site at http://wow.nrri.umn.edu/wow/.

5.2.5   Monitoring Your Sound (MY Sound) (Long Island Sound, New York)

Introduction

The MY Sound project provides real-time water quality monitoring data from Long Island Sound to a broad spectrum of users, including government, academia, industry, organizations, and the general public. The project recognizes that water quality in Long Island Sound is an issue that affects everyone, not just those who live along the coast. If water quality is poor, the value of the Sound as an economic, recreational, and natural resource decreases; if water quality is good, people use it and it is a vital resource. A major goal of the project is to enhance and broaden the user's appreciation, knowledge, and use of Long Island Sound. The project, which was coordinated by a stakeholder committee comprised of project partners and stakeholder representatives, uses the Internet, local media, information kiosks, orientation briefings, and printed material.

The project has established five water quality monitoring stations near New London and Bridgeport Harbors. The EMPACT focus areas include Bridgeport

Harbor and the greater CT-NY-Long Island metropolitan area. The monitoring stations collect data for the following parameters:

- Water temperature
- Conductivity/salinity
- Transmissivity
- Dissolved oxygen
- Nutrients/nitrate
- Chlorophyll
- Surface hydrocarbons
- Current speed and direction
- Selected meteorological parameters

Lessons, Tools, and Activities

At the time of publishing this handbook, the MY Sound project was developing curriculum support tools that can be used by teachers of environmental science, physics, and math courses. The materials will be geared toward students in grades 8 through 12. Specific components under development include:

- Fact sheets on topics related to the environmental health of Long Island Sound.
- Student exercises that use time series and statistical data on Long Island Sound phenomena to illustrate science and math principles and enhance knowledge of the Sound.
- Guided Internet explorations that lead teachers and students through key Web sites to investigate marine science topics.

Examples of future student exercises include:

- A Long Island Sound lobster mortality exercise that illustrates the use of statistics in investigating lobster population decline in recent years (will involve both manual calculations and spreadsheet development).
- A sunken oil barge salvage exercise that illustrates hydrodynamic principals important in re-floating a sunken oil barge in eastern Long Island Sound.
- A small boat drift exercise using MY Sound wind and current data that illustrates the use of vector addition in conducting a search and rescue operation.
- An ocean data analysis exercise using wind and dissolved oxygen time series data that illustrate the concepts of hypoxia, temperature stratification, and vertical mixing on a Summer 2000 event in western Long Island Sound.

Examples of guided Internet investigations include:

- Waste water pollution (municipal and industrial)

- Oil and hazardous chemical spills

- Non-point source pollution

- Invasive species

- Marine debris

- Habitat modification and restoration

Resources

For additional information on the MY Sound project and status of the curriculum component, contact Pete Tebeau at 860 446-0193 or visit the MY Sound Web site at <www.MYSound.uconn.edu>.

5.2.6 Online Dynamic Watershed Atlas (Seminole County, Florida)

Introduction

The Seminole County Watershed Atlas is designed to provide citizens, scientists, and planners of the Seminole County region with comprehensive and current water quality, hydrologic, and ecological data, as well as a library of scientific and educational resources on ecology and management. The Atlas was created to provide a "one stop information shop" for concerned citizens and scientists who live and work on water bodies and have found it difficult to gather the information they need from the many agencies that collect the related data. The Atlas functions as a warehouse for a variety of water resources information, including documents and educational links. The Atlas also is a rich resource that educates citizens about the data presented and gives scientists easy access to the specialized information they need.

Lessons, Tools, and Activities

As part of the Atlas project, Seminole County initiated a water quality and hydrology curriculum component in September 2001. The curriculum, which is being developed in conjunction with the University of South Florida and the Seminole County School Board, along with several other minor partners, is expected to be completed by January 2004. Designed for grades 5 through 12, the curriculum will be provided to county schools, a local environmental studies center, and other interested environmental education groups. The curriculum will cross several disciplines, including math, science, and social studies. Project organizers are expecting that in the future, other counties will develop their own watershed databases and could adapt the Seminole County curriculum to meet their needs.

Teachers will work with county staff to design the curriculum and will then train other teachers how to use it. Curriculum staff will develop both teacher and student guides. Teachers and students will need Internet access to use the curriculum, and optional field activities are under consideration, which might require environmental monitoring equipment.

Resources

For additional information on the Seminole County Watershed Atlas project or curriculum, contact Kim Ornberg at 407 665-5738 or visit the project Web site at <www.seminole.wateratlas.usf.edu>.

5.2.7   Onondaga Lake/Seneca River (Syracuse, New York)

Introduction

The Onondaga Lake/Seneca River EMPACT project provides environmental information on the health of the Onondaga Lake/Seneca River ecosystems to students, researchers, and the local Syracuse community. Onondaga Lake is one of the most polluted lakes in the United States, with fishing and swimming prohibited and several water quality standards routinely violated. The lake pollution affects adjoining waterways, including the Seneca River. In 1998, local, state, and federal authorities agreed on a 15-year staged program to address the impacts of sewage pollution on the lake and river, and in 1999, the project was awarded an EMPACT grant. The program, a partnership between the Syracuse City School District, the Upstate Freshwater Institute, State University of New York–School of Environmental Science and Forestry, Syracuse University, and local businesses, collects and delivers critical near real-time data from remote underwater sample stations, or RUSS units, in the lake and river. The goals of the project include:

■ Applying and advancing innovative remote monitoring technology to meet the acute present and future monitoring needs for the lake and river.

■ Addressing the community's lack of understanding concerning the degraded conditions of the ecosystems.

■ Promoting excellence in teaching, learning, and research.

The lasting benefits of the projects will include:

■ Addition of critical capabilities to the long-term monitoring program.

■ Creation of vehicles to communicate important characteristics and findings to all stakeholders.

■ A community that is more engaged in critical environmental decision-making.

Lessons, Tools, and Activities

Three educational resources have been developed to support classroom instruction and connect school curricula to the Onondaga Lake-Seneca River EMPACT Project. Grade-level course guides for early primary (K–3), elementary (4–6), intermediate (7–9), and commencement (10–12) students have been developed to supplement project efforts. The lessons in the guides were designed to be implemented as part of a regular science course. For example, students could learn weather principles by studying the RUSS meteorological data. There are some teachers who are using the materials in special Onondaga Lake Units. These types of units are taught in the spring and review all the concepts of a course.

Several essential understandings form the basis of the course guides. A committee of teachers representing all grade levels and content areas of the Syracuse City School District analyzed the issues and concepts impacting Onondaga Lake and its watershed. Through their analysis, they identified the following essential understandings:

- Several dynamic processes are constantly reshaping the Onondaga Lake Watershed, including:

    - Succession: The continuing process in which an ecosystem evolves to maximize the cycling and utilization of resources.

    - Seasonal changes: The processes involved with the motion of the earth and moon about the sun, and the processes that occur in response to their motion.

    - Human processes: The processes involved with human activity and the environmental impacts that result.

- The earth is a closed system.

    - Life is sustained by and is part of a set of cyclic processes.

    - All resources used by humans were developed through a series of cyclic processes.

    - All waste products, if not transformed, will remain in the global system.

- Humans make decisions. Human action is directed primarily by thought and decisionmaking in an effort to improve the quality of life.

- Efficient and effective communication skills are necessary for success at any task or performance.

In addition to the essential understandings that were developed under the project, teachers developed essential questions to drive classroom inquiry and research. The primary question to drive inquiry in all classrooms and content areas is "How do we make the decisions necessary to develop and maintain a healthy community?" The Onondaga Lake and Seneca River are two components of the watershed ecosystem. Because all components of the ecosystem are interconnected, monitoring changes in water quality provides insight into the overall health of the watershed and the communities it supports. As a result, students are challenged to assess their community and their impact upon it. The key questions for driving inquiry for each essential understanding of the project are:

- What are the processes that impact our community?

- How does material enter and leave our community?

- What happens to these materials when they interact with our community?

- How do these materials impact upon and/or affect our community?

- How do humans, individually and in groups, make decisions?

- How do people make the decisions necessary to communicate effectively with each other?

For each grade level, there are lessons covering each essential understanding and key question. For example, to address the key understanding of dynamic processes and the key question, "What are the processes that impact our community?" the Onondaga Lake curriculum includes the lesson "Shake, Rattle, and Role: Earth's Dynamic Processes." The theme, topics, and project work vary by grade level. As an example, for 10th grade, the theme of the lesson is cycles and cyclic processes; the curriculum topics include biological interactions with dynamic changes, lake biology, and Onondaga Creek Watershed ecology; students assume the role of research botanists, microbiologists, zoologists, entomologists, and environmental engineers and present a physical model as a project.

Resources

For more information on the Onondaga Lake/Seneca River project, contact Richard List at 315 435-5842 or at <rlist@freeside.scsd.k12.ny.us> and visit the project Web site at <www.ourlake.org>.

6.0 Land-Use and Soil-Based Projects

6.1 Teacher Tips

Soil is a dynamic resource that supports plant life. It is comprised of a number of different materials, including sand, silt, clay, organic matter, and many species of living organisms. Therefore, soil has biological, chemical, and physical properties, some of which can change depending on how the soil is managed. The Soil Science Society of America defines soil quality as "the capacity of a specific kind of soil to function, within natural or managed ecosystem boundaries, to sustain plant and animal productivity, maintain or enhance water and air quality, and support human health and habitation." Management that enhances soil quality benefits cropland, rangeland, and woodland productivity. In addition, enhanced soil quality benefits water quality, air quality, and wildlife habitat. Soil provides several essential services or functions:

EPA's Office of Research and Development (ORD) conducts research in innovative monitoring and measurement technologies, as well as in tools to interpret data streams and to increase the quality and the number of environmental parameters that can be monitored and reported in EMPACT communities. Although there are currently no research grants researching soil monitoring technologies, teaching students about soil quality is important, so this handbook provides background information as a resource for the teacher.

- Soil supports the growth and diversity of plants and animals by providing a physical, chemical, and biological environment for the exchange of water, nutrients, energy, and air.

- Soil regulates the distribution of rain or irrigation water between infiltration and runoff, and it regulates the flow and storage of water and the materials found in it, such as nitrogen, phosphorus, pesticides, and nutrients.

- Soil stores, moderates the release of, and cycles plant nutrients and other elements.

- Soil acts as a filter to protect the quality of water, air, and other resources.

Soil quality is evaluated using indicators that reflect changes in the capacity of the soil to function. Useful indicators are those that are sensitive to change and that change in response to management. Some examples include soil erosion, sediment deposition, soil biodiversity, water capacity, and pesticides. Monitoring of soil quality indicators over time identifies changes or trends in the functionality or quality of the soil. Monitoring can be used to determine the success of management practices or the need for changes or adjustments.

Most soil-related EMPACT projects focus either on lead exposure from residential soils or the status of brownfield properties. (See <www.epa.gov/empact/soil.htm> for more information on these projects, which do not currently have curriculum components.)

Another topic associated with soil is land use and urban sprawl. Urban sprawl can be defined as the unplanned, unlimited extension of neighborhoods outside of a city's limits, usually associated with low density residential and commercial settlements, dominance of transportation by automobiles, and widespread strip commercial development. Over the past 50 years, American cities have been experiencing an accelerated urbanization and suburbanization process resulting from rapid technological advancement and relatively steady economic growth. Some argue that urban sprawl leads to inefficient land use patterns. Communities can implement a number of growth management programs to encourage more efficient and environmentally sound development patterns.

6.2   The Tools

6.2.1   Northeast Ohio Urban Growth Simulator

Introduction

The Northeast Ohio (NEO) EMPACT project compiled urban sprawl data to create a land-use computer modeling tool. Developed by Kent State University, Cleveland State University, and the University of Akron, it provides citizens with local urban sprawl information and development scenarios for Northeast-Ohio. This information helps decision-making on how the region should grow and provides possible land use consequences that might arise from different kinds of growth (i.e., farmland loss, wetland destruction).

Lessons, Tools, and Activities

As part of the NEO EMPACT project, a handbook was developed to introduce teachers and students to the importance of understanding urban sprawl in their communities. Urban Sprawl in Northeast Ohio is arranged in thematic and developmental order to provide students with a comprehensive understanding of urban sprawl and its effects on the environment.

The handbook for educators and students includes detailed background information, lessons, and activities focused on urban sprawl. It progresses from developing an understanding of urban sprawl to discussing concrete actions students and teachers can take to raise awareness of urban sprawl. The major sections of the handbook include:

■   The introductory section, *All About Urban Sprawl—Notes for Educators,* provides detailed background information on urban sprawl and how it relates to other environmental problems such as air and water pollution and acid rain.

■   The 10 *Experiments and Exercises* on urban sprawl provide hands-on lessons in urban sprawl. Geared towards specific grades, the experiments and exercises cover land use planning, various types of air pollution (e.g., particulates, carbon dioxide), soil buffering, air quality as it relates to combustion byproducts, habitat destruction, water pollution, and city planning.

- The *Students, Urban Sprawl, and the Internet* section is complemented by the online *Urban Growth Simulator* and its *Self-Guided Workbook*, which allow students to simulate how their community would change with future development. The workbook describes the Urban Growth Simulator Web site, and includes four guided simulation exercises.

- The *Urban Sprawl Activities for Younger Students* focus on developing students' oral, visual, and writing skills. Activities include conducting a mock interview with a famous environmentalist, a word search and crossword puzzle, writing an urban sprawl bill, determining the authority of various levels of government (i.e., federal, state, local) to pass land use laws, and designing urban sprawl posters for display in the community.

- *Urban Sprawl—What Students Can Do* offers teachers and students suggestions for reducing sprawl and its side effects in the local community and at school.

- *Urban Sprawl World Wide Web Resources for Educators* lists sources of additional information on urban sprawl for educators and students.

Resources

For additional information on the NEO Urban Sprawl curriculum handbook, contact Adam Zeller of the Earth Day Coalition at 216 281-6468 or <azeller@earthdaycoalition.org> and visit the NEO EMPACT Web site at <http://empact.nhlink.net>.

Appendix A: Additional Resources

U.S. Environmental Protection Agency (EPA) Office of Solid Waste

<www.epa.gov/epaoswer/osw/teacher.htm>

This Web site provides educational tools and a list of related publications, including:

- *Let's Reduce and Recycle: Curriculum for Solid Waste Awareness*
- *School Recycling Programs: A Handbook for Educators*
- *Adventures of the Garbage Gremlin*

EPA also lists a wealth of activities including the "Planet Protectors Coloring Book."

The Globe Program: Global Learning and Observations To Benefit the Environment

<www.globe.gov>

This Web site provides science and education resources including teacher guides, workshops, and tools, such as a geography quiz and cloud identification quiz.

Natural Resource Conservation Service

<www.nrcs.usda.gov/feature/education>

This Web site includes ideas and educational tools for teachers.

U.S. Department of Agriculture (USDA) for Kids

<www.usda.gov/news/usdakids/index.html>

"USDA for Kids" Web site is a great resource for educational tools, including a food pyramid guide, Smokey the Bear, and "Food for Thought."

National Soil Survey Center

<www.statlab.iastate.edu/soils/nssc>

This Web site provides information on soil science education.

National Geographic Society

<www.nationalgeographic.com>

This Web site provides extensive teacher resources related to geography and science.

North American Association of Environmental Education (NAAEE)

1255 23rd Street NW., Suite 400
Washington, DC 20037-1199
202 884-8912
Fax: 202 884-8701
<www.naaee.org>

NAAEE was established in 1971 as a network of professionals and students working in environmental education. NAAEE's members are located throughout North America and in more than 40 countries around the world; they believe that education is the key to ensuring a healthy, sustainable environment and improving the quality of life on earth. Members can join various sections: Elementary and Secondary Education, College and University Environmental Programs, and Non-formal Education.

Association for Supervision and Curriculum Development (ASCD)

1250 North Pitt Street
Alexandria, VA 22314
703 549-9110
<www.ascd.org>

ASCD, an education association, serves its members through publications, professional development opportunities, research and information searches, the Curriculum and Technology Resource Center, and affiliates in each state and several foreign countries. Resources include information on staff development practices, cooperative learning, peer coaching, and science and social studies content for schools.

National Science Teachers Association (NSTA)

1840 Wilson Boulevard
Arlington, VA 22201-3000
703 243-7100
Fax: 703 243-7177
<www.nsta.org>

The National Science Teachers Association (NSTA) is committed to improving science education at all levels, preschool through college. NSTA produces several publications, conducts national and regional conventions, and provides scholarships, teacher-training workshops, educational tours, and an employment registry. The Web site provides an extensive range of resources for teachers of students of all levels; journals and books on science education and instruction are also available.

National School Boards Association (NSBA)

1680 Duke Street
Alexandria, VA 22314
703 838-6722
<www.nsba.org>

The National School Boards Association is a national federation of state school boards. NSBA produces "Electronic School," a free online technology publication for K-12 educators. NSBA houses the Institute for the Transfer of Technology to Education (ITTE), a program to help advance the wise use of technology in public education.

Appendix B: Glossary of Terms

Air Terms

Acid rain: Air pollution produced when acid compounds formed in the atmosphere are incorporated into rain, snow, fog, or mist. The acid compounds come from sulfur oxides and nitrogen oxides, products of burning coal and other fuels and from certain industrial processes. Acid rain can impact the environment and human health and damage property.

Atmosphere: A thin layer of gases surrounding the Earth, composed of 78 percent nitrogen, 21 percent oxygen, 0.9 percent argon, 0.03 percent carbon dioxide, and trace amounts of other gases. There is no exact place where the atmosphere ends; it just gets thinner and thinner, until it merges with outer space.

Basal cell carcinoma: Skin cancer tumors that might appear as slow-growing, translucent, pearly nodules, which might crust, discharge pus, or even bleed. These tumors typically develop where you are most exposed to the sun—on the face, lips, tops of ears, and hands.

Carbon monoxide (CO): A colorless, odorless, poisonous gas produced by the incomplete burning of solid, liquid, and gaseous fuels. Appliances fueled with natural gas, liquified petroleum (LP gas), oil, kerosene, coal, or wood may produce CO. Burning charcoal produces CO and car exhaust contains CO.

Chlorofluorocarbons (CFCs): Stable, low toxic, and inexpensive chemicals that were most commonly used as refrigerants, solvents, and aerosol propellants. CFCs and their relatives, when released into the air, rise into the stratosphere and take part in chemical reactions that result in reduction or depletion of the stratospheric ozone layer. The 1990 Clean Air Act includes provisions for reducing releases (emissions) and eliminating production and use of these ozone-destroying chemicals.

Clean Air Act: The original Clean Air Act was passed in 1963, but our national air pollution control program is actually based on the 1970 version of the law. The 1990 Clean Air Act Amendments are the most far-reaching revisions of the 1970 law.

Criteria air pollutants: A group of very common air pollutants regulated by EPA on the basis of criteria (information on health and/or environmental effects of pollution).

Emission: Release of pollutants into the air from a source. Continuous emission monitoring systems (CEMS) are machines that some large sources are required to install, to make continuous measurements of pollutant release.

EMPACT: Environmental Monitoring for Public Access and Community Tracking, a program begun by EPA in 1997, helps communities collect, manage, and distribute environmental information, providing residents with up-to-date and easy-to-understand information they can use to make informed, day-to-day decisions.

Greenhouse effect: A natural phenomenon whereby clouds and greenhouse gases, such as water vapor and carbon dioxide, trap some of the Sun's heat in the atmosphere. The greenhouse effect helps regulate the temperature of the Earth. Human activities are adding greenhouse gases to the natural mix.

Greenhouse gases: Human activities, such as fuel burning, are adding greenhouse gases to the atmosphere. Because these gases remain in the atmosphere for decades to centuries (depending on the gas) global temperatures will rise.

Melanoma: The most fatal form of skin cancer. Malignant melanomas may appear suddenly without warning as a dark mole or other dark spot on the skin and can spread quickly.

Monitoring (monitor): Measurement of air pollution is referred to as monitoring. Continuous emission monitoring systems (CEMS) will measure, on a continuous basis, how much pollution is being released into the air. The 1990 Clean Air Act requires states to monitor community air in polluted areas to check on whether the areas are being cleaned up according to schedules set out in the law.

Nitrogen oxides (NOx): A criteria air pollutant. Nitrogen oxides are produced from burning fuels, including gasoline and coal, and react with volatile organic compounds to form smog. Nitrogen oxides are also major components of acid rain.

Ozone (O3): An ozone molecule consists of three oxygen atoms. Stratospheric ozone shields the Earth against harmful rays from the sun, particularly ultraviolet B. Ground-level ozone contributes to smog.

Ozone depletion: The ozone layer is damaged when substances such as chlorofluorocarbons accelerate the natural process of destroying and regenerating stratospheric ozone. As the ozone layer breaks down, it absorbs smaller amounts of UV radiation, allowing more of it to reach the earth.

Particulates, particulate matter: A criteria air pollutant. Particulate matter includes dust, soot, and other tiny bits of solid materials that are released into and move around in the air.

Pollutants (pollution): Unwanted chemicals or other materials found in the air.

Smog: A mixture of pollutants, principally ground-level ozone, produced by chemical reactions in the air involving smog-forming chemicals. A major portion of smog-formers come from burning of petroleum- based fuels such as gasoline. Major smog occurrences are often linked to heavy motor vehicle traffic, sunshine, high temperatures, and calm winds or temperature inversion (weather condition in which warm air is trapped close to the ground instead of rising).

Source: Any place or object from which pollutants are released.

Spectrophotometer: An instrument for measuring the relative intensities of light in different parts of the spectrum used to measure the amount of UV radiation reaching the earth.

Squamous cell carcinoma: Skin cancer tumors that might appear as nodules or red, scaly patches, which can develop into large masses and spread to other parts of the body.

Stratosphere: The stratosphere starts just above the troposphere and extends to 50 kilometers (31 miles) high. The temperature in this region increases gradually to -3 degrees Celsius, due to the absorption of ultraviolet radiation. The ozone layer, which absorbs and scatters the solar ultraviolet radiation, is in this layer. Ninety-nine percent of air is located in the troposphere and stratosphere.

Stratospheric ozone: A bluish gas composed of three oxygen atoms. Natural processes destroy and regenerate ozone in the atmosphere. When ozone-depleting substances such as chlorofluorocarbons accelerate the destruction of ozone, there is less ozone to block UV radiation from the sun, allowing more UV radiation to reach the earth.

Sulfur dioxide: A criteria air pollutant. Sulfur dioxide is a gas produced by burning coal, most notably in power plants. Sulfur dioxide plays an important role in the production of acid rain.

Sunscreen: A substance, usually a lotion, that you can apply to protect your skin from UV radiation. It works by reflecting UV radiation away from your skin in addition to absorbing UV radiation before it can penetrate your skin.

SunWise School Program: EMPACT program that aims to teach grades K-8 school children and their caregivers how to protect themselves from overexposure to the sun. The program raises children's awareness of stratospheric ozone depletion and ultraviolet radiation and encourages simple sun safety practices.

Troposphere: The troposphere is the lowest region in the Earth's (or any planet's) atmosphere, starting at ground (or water) level up and reaching up to about 11 miles (17 kilometers) high. The weather and clouds occur in the troposphere.

Ultraviolet B (UVB): A type of sunlight. Ultraviolet B exposure has been associated with skin cancer, eye cataracts, and damage to the environment. The ozone in the stratosphere, high above the Earth, filters out ultraviolet B rays and keeps them from reaching the Earth. Thinning of the ozone layer in the stratosphere results in increased amounts of ultraviolet B reaching the Earth.

UV Index: A tool developed by the National Weather Service that predicts the next day's UV intensity on a scale from 0 to 10+, helping people determine appropriate sun-protective behaviors.

UV radiation: A portion of the electromagnetic spectrum with wavelengths shorter than visible light. UV radiation produced by the sun is responsible for sunburn and other adverse health effects. Scientists classify UV radiation into three types: UVA, UVB, and UVC.

Volatile organic compounds (VOCs): Chemicals that produce vapors readily at room temperature and normal atmospheric pressure, so that vapors escape easily from volatile liquid chemicals. Organic chemicals all contain the element carbon and are the basic chemicals found both in living things and in products derived from living things, such as coal, petroleum and refined petroleum products. Many volatile organic chemicals are also hazardous air pollutants.

Water Terms

Abiotic: Not alive; non-biological. For example, temperature and mixing are abiotic factors that influence the oxygen content of lake water, whereas photosynthesis and respiration are biotic factors that affect oxygen solubility.

Acid: A solution that is a proton (H+) donor and has a pH less than 7 on a scale of 0-14. The lower the pH the greater the acidity of the solution.

Acidity: A measure of how acidic a solution may be. A solution with a pH of less than 7.0 is considered acidic. Solutions with a pH of less than 4.5 contain mineral acidity (due to strong inorganic acids), while a solution having a pH greater than 8.3 contains no acidity.

Acid rain: Precipitation having a pH lower than the natural range of ~5.2 - 5.6; caused by sulfur and nitrogen acids derived from human-produced emissions.

Acidification: The process by which acids are added to a water body, causing a decrease in its buffering capacity (also referred to as alkalinity or acid neutralizing capacity), and ultimately a significant decrease in pH that may lead to the water body becoming acidic (pH < 7).

Algae: Simple single-celled, colonial, or multi-celled aquatic plants. Aquatic algae are (mostly) microscopic plants that contain chlorophyll and grow by photosynthesis and lack roots, stems (non- vascular), and leaves.

Alkalinity: Acid neutralizing or buffering capacity of water; a measure of the ability of water to resist changes in pH caused by the addition of acids or bases. Therefore, it is the main indicator of susceptibility to acid rain. A solution having a pH below about 5 contains no alkalinity.

Anoxia: Condition of being without dissolved oxygen.

Anthropogenic: A condition resulting from human activities.

Aquatic respiration: Refers to the use of oxygen in an aquatic system, including the decomposition of organic matter and the use of oxygen by fish, algae, zooplankton, aquatic macrophytes, and microorganisms for metabolism.

Base: A substance which accepts protons (H+) and has a pH greater than 7 on a scale of 0-14; also referred to as an alkaline substance.

Basin: Geographic land area draining into a lake or river; also referred to as drainage basin or watershed.

Benthic: Refers to being on the bottom of a lake.

Bioaccumulation: The increase of a chemical's concentration in organisms that reside in environments contaminated with low concentrations of various organic compounds. Also used to describe the progressive increase in the amount of a chemical in an organism resulting from rates of absorption of a substance in excess of its metabolism and excretion. Certain chemicals, such as PCBs, mercury, and some pesticides, can be concentrated from very low levels in the water to toxic levels in animals through this process.

Biochemical oxygen demand (BOD): Sometimes referred to as Biological Oxygen Demand (BOD). A measure of the amount of oxygen removed (respired) from aquatic environments by aerobic microorganisms either in the water column or in the sediments. Primarily of concern in wastewater "streams" or systems impacted by organic pollution.

Biomass: The weight of a living organism or group of organisms.

Biotic: Referring to a live organism; see abiotic.

Buffer: A substance that tends to keep pH levels fairly constant when acids or bases are added.

Chlorophyll: Green pigment in plants that transforms light energy into chemical energy during photosynthesis.

Clarity: Transparency; routinely estimated by the depth at which you can no longer see a Secchi disk. The Secchi disk, an 8-inch diameter, weighted metal plate, is lowered into water until it disappears from view. It is then raised until just visible. An average of the two depths, taken from the shaded side of the boat, is recorded as the Secchi depth.

Conductivity (electrical conductivity and specific conductance): Measures water's ability to conduct an electric current and is directly related to the total dissolved salts (ions) in the water. Called EC for electrical conductivity, it is temperature-sensitive and increases with higher temperature.

Dissolved oxygen (DO or O2): The concentration of free (not chemically combined) molecular oxygen (a gas) dissolved in water, usually expressed in milligrams per liter, parts per million, or percent of saturation. Adequate concentrations of dissolved oxygen are necessary for the life of fish and other aquatic organisms.

Dissolved solids concentration: The total mass of dissolved mineral constituents or chemical compounds in water; they form the residue that remains after evaporation and drying.

Ecosystem: All of the interacting organisms in a defined space in association with their interrelated physical and chemical environment.

Epilimnion: The upper, wind-mixed layer of a thermally stratified lake. This water is turbulently mixed at some point during the day, and, because of its exposure, can freely exchange dissolved gases (such as O2 and CO2) with the atmosphere.

Eutrophication: Unhealthy increases in the growth of phytoplankton. Symptoms of eutrophication include algal blooms, reduced water clarity, periods of hypoxia, and a shift toward species adapted toward these conditions.

Evaporation: The process of converting liquid to vapor.

Food chain: The transfer of food energy from plants through herbivores to carnivores. For example, algae are eaten by zooplankton, which in turn are eaten by small fish, which are then eaten by larger fish, and eventually by people or other predators.

Food web: Food chains connected into a complex web.

Hydrogen: Colorless, odorless, and tasteless gas; combines with oxygen to form water.

Hydrology: The study of water's properties, distribution, and circulation on Earth.

Hypolimnion: The bottom and most dense layer of a stratified lake. It is typically the coldest layer in the summer and warmest in the winter. It is isolated from wind mixing and typically too dark for much plant photosynthesis to occur.

Hypoxia: A deficiency of oxygen reaching the tissues of the body.

Isothermal: Constant in temperature.

Leach: To remove soluble or other constituents from a medium by the action of a percolating liquid, as in leaching salts from the soil by the application of water.

Metalimnion: The middle or transitional zone between the well-mixed epilimnion and the colder hypolimnion layers in a stratified lake.

Nonpoint source: Diffuse source of pollutant(s); not discharged from a pipe; associated with land use such as agriculture, contaminated groundwater flow, or onsite septic systems.

Nutrient loading: Discharging of nutrients from the watershed (basin) into a receiving water body (lake, stream, wetland).

Oxygen: An odorless, colorless gas; combines with hydrogen to form water; essential for aerobic respiration. See respiration.

Oxygen solubility: The ability of oxygen gas to dissolve into water.

Parameter: Whatever it is you measure; a particular physical, chemical, or biological property that is being measured.

pH: A measure of the concentration of hydrogen ions.

Phosphorus: Key nutrient influencing plant growth in lakes.

Photosynthesis: The process by which green plants convert carbon dioxide (CO_2) dissolved in water to sugars and oxygen using sunlight for energy. Photosynthesis is essential in producing a lake's food base and is an important source of oxygen for many lakes.

Phytoplankton: Microscopic floating plants, mainly algae, that live suspended in bodies of water and that drift about because they cannot move by themselves or because they are too small or too weak to swim effectively against a current.

Respiration: The metabolic process by which organic carbon molecules are oxidized to carbon dioxide and water with a net release of energy.

Solubility: The ability of a substance to dissolve into another.

Solution: A homogenous mixture of two substances.

Solvent: A substance that has the ability to dissolve another.

Stormwater discharge: Precipitation and snowmelt runoff (e.g., from roadways, parking lots, roof drains) that is collected in gutters and drains; a major source of nonpoint source pollution to water bodies.

Temperature: A measure of whether a substance is hot or cold.

Total Dissolved Solids (TDS): The amount of dissolved substances, such as salts or minerals, in water remaining after evaporating the water and weighing the residue.

Turbidity: Degree to which light is blocked because water is muddy or cloudy.

Turnover: Fall cooling and spring warming of surface water make density uniform throughout the water column, allowing wind and wave action to mix the entire lake. As a result, bottom waters contact the atmosphere, raising the water's oxygen content.

Water Column: A conceptual column of water from lake surface to bottom sediments.

Watershed: All land and water areas that drain toward a river or lake; also called a drainage basin or water basin.

Soil Terms

Bedrock: Consolidated rock.

Brownfields: Abandoned, idled, or underused industrial and commercial facilities where expansion or redevelopment is complicated by real or perceived environmental contamination.

Clay: Soil composed mainly of fine particles of hydrous aluminum silicates and other minerals. Soil composed chiefly of this material has particles less than a specified size.

Erosion: The wearing away of the land surface by running water, wind, ice, other geological agents, or human activity.

Infiltration: The downward entry of water through the soil surface.

Limestone: A white to gray, fine-grained rock made of calcium carbonate.

Percolation: Water that moves through the soil at a depth below the root zone.

Sand: A loose granular material that results from the disintegration of rocks. It consists of particles smaller than gravel but coarser than silt

Sandstone: A very grainy rock that comes in many colors, including gray, red, or tan.

Sedimentary rock: Rock that has formed from compressed sediment, like sand, mud, and small pieces of rocks.

Shale: Dark-colored rock that is usually black, deep red, or gray-green. It has a fine grain and is usually found below sandstone, not on the surface. Shale was formed from fine silt and clay.

Silt: Predominantly quartz mineral particles that are between the size of sand and clay in diameter. Silt, like clay and sand, is a product of the weathering and decomposition of preexisting rock.

Soil: Soil is made up of minerals (rock, sand, clay, silt), air, water, and organic (plant and animal) material. There are many different types of soils, and each one has unique characteristics, like color, texture, structure, and mineral content.

Soil contamination: Pollution caused by a number of activities, including the dumping of hazardous substances, pesticide and fertilizer use, and industrial or chemical processes. Pollutants in soils can also be transported to groundwater sources and into the air. Contaminated soils are often a major concern at brownfield and Superfund sites. Common soil contaminants include arsenic, benzene, cyanide, lead, and mercury.

Soil formation: Soil is formed slowly as rock erodes into tiny pieces near the Earth's surface. Organic matter decays and mixes with rock particles, minerals, and water to form soil.

Soil texture: Distribution of individual particles of soil.

Soil washing: A technology that uses liquids (usually water, sometimes combined with chemical additives) and a mechanical process to scrub soils of contaminants.

Superfund: The Federal government's program to clean up the nation's uncontrolled hazardous waste sites.

Topsoil: Soil consisting of a mixture of sand, silt, clay, and organic matter. Topsoil is rich in nutrients and supports plant growth.

Urban sprawl: The unplanned, unlimited extension of neighborhoods outside of a city's limits, usually associated with low density residential and commercial settlements, dominance of transportation by automobiles, and widespread strip commercial development.

Appendix C: Activities by Grade Level

Curriculum							Grade							
	K	1	2	3	4	5	6	7	8	9	10	11	12	12+
Airbeat					X	X	X	X	X	X	X	X	X	
Air Currents							X	X	X	X	X	X	X	
Air Info Now: Environmental Monitoring for Public Access and Community Tracking					X	X	X	X	X	X	X	X	X	
AIRNow			X	X	X	X								
Boulder Area Sustainability Information Network					X	X	X	X	X	X	X	X	X	
Burlington Eco-Info							X	X	X	X	X	X	X	
Community Accessible Air Quality Monitoring Assessment (Northeast Ohio)					X	X	X	X	X					
ECOPLEX	X	X	X	X	X	X	X	X	X					
Lake Access (WOW)												X	X	X
Monitoring Your Sound									X	X	X	X	X	
Online Dynamic Watershed Atlas (Seminole County, FL)						X	X	X	X	X	X	X	X	
Onondaga Lake/Seneca River	X	X	X	X	X	X	X	X	X	X	X	X	X	
Northeast Ohio Urban Growth Simulator					X	X	X	X	X					

Appendix D: Activities by Subject

Curriculum	Math	Language Arts	Subject Science	Social Studies	Art
Airbeat	X	X	X		
Air CURRENTS	X	X	X	X	X
Air Info Now: Environmental Monitoring for Public Access and Community Tracking		X	X		
AIRNow	X		X		X
Boulder Area Sustainability Information Network			X	X	
Burlington Eco-Info			X		
Community Accessible Air Quality Monitoring Assessment (Northeast Ohio)		X	X		X
ECOPLEX	X		X		X
Lake Access	X		X		
Monitoring Your Sound	X		X		
Online Dynamic Watershed Atlas (Seminole County, FL)	X		X	X	
Onondaga Lake/Seneca River			X		
Northeast Ohio Urban Growth Simulator			X	X	

Appendix E: Selected Lesson Plans and Activities

AirInfo Now

- Group Details – Blue Group: Weather (PDF)
  Data Sheet – Blue Group: Weather (Excel)
- Group Details – Brown Group: Visibility (PDF)
  Data Sheet – Brown Group: Visibility (Excel)
- A Guide to CO-City (PDF)
- So What's Making it Look Brown Outside? Collecting and Measuring Particulate Matter (PDF)
- What's the Connection Between Convection and Inversion? Convection Currents and Temperature Inversion (PDF)
- Getting a Handle on Greenhouse Gases: Your Family's Impact on the Greenhouse Effect (PDF)
- Helping to Find a Solution to Air Pollution! (PDF)
- Green Group: Location (PDF)
  Green Group: Location (Excel)
- Real-Time Air Quality Activity: Groups (PDF)
- Practice Data Sheet (Excel)
- Group Details – Red Group: Time (PDF)
  Data Sheet – Red Group: Time (Excel)
- Real-Time Air Quality Activity: Student Sheets(PDF)
  Real-Time Air Quality Activity: Teacher Sheets(PDF)
- Group Details – Yellow Group: Health (PDF)
  Data Sheet – Yellow Group: Health (Excel)

Airnow

- Air Quality Index Poster: Are you breathing clean air? (PDF)
- Air Quality Index: A Guide to Air Quality and Your Health (PDF)
- Air Quality Index Kids Website: Teacher's Reference (PDF)
- Green Day Poster (PDF)
- Orange Day Poster (PDF)
- Air Quality Index Posters (PDF)
- Purple Day Poster (PDF)
- Red Day Poster (PDF)
- Yellow Day Poster (PDF)

ECOPLEX

- **UV**
  - UV/7-2: Spotlight the Sun Data Table (PDF)
  - Ozone Chemistry: Formation & Depletion(PDF)
  - 8th Grade Lesson Plan – UV: Chemistry of Ozone Depletion(PDF)
  - 5th Grade Lesson Plan – UV: Check It Out! (PDF)
  - First Grade UV: Catching and Counting UV Rays! (PDF)
  - 4th Grade UV Lesson: What Depletes Our Ozone? Me and My Zone! (PDF)
  - Kindergarten UV: UV and Me! (PDF)
  - Second Grade UV: The Air Out There – UV and Ozone (PDF)

- UV/7-1: Distribution of the Sun's Rays (PDF)
    - 6th Grade UV: Friend or Foe (PDF)
    - 3rd Grade UV Lesson: When Good Ozone Goes Bad (PDF)

- **Water Quality**
    - Third Grade Water Quality: Test, Test, Is This Water Safe? (PDF)
    - Fourth Grade Water Quality: Chain, Chain, Chain, Chain of Food (PDF)
    - Fifth Grade Water Quality: Tick Tock Toxins (PDF)
    - 6th Grade Water Quality Lesson: Water O2 and You! (PDF)
    - 7th Grade Water Quality Lesson: Taxa-Rich and Taxa-Poor! (PDF)
    - Water Quality 1-1 Record Sheet (PDF)
    - Water Quality 2-1 Record Sheet (PDF)
    - Water Quality 4-1 Datasheet (PDF)
    - Water Quality 5-1 Datasheet (PDF)
    - First Grade Water Quality: Water – It's a Gas…Sometimes! (PDF)
    - Kindergarten Water Quality: Water in Me (PDF)
    - Second Grade Water Quality: Amazing Water (PDF)

- **Water Quantity**
    - 7th Grade Water Quantity: Water Use and Abuse (PDF)
    - 3rd Grade Water Quantity: Name That Surface Water (PDF)
    - 4th Grade Water Quantity: H2O is Underground Too! (PDF)
    - 5th Grade Water Quantity: What-A-Shed (PDF)
    - WQT/6-1: Water vs. Land and Sea (PDF)
    - WQT/6-2: Diagram for Stream Table (PDF)
    - 6th Grade Water Quantity: The Ups and Downs of Your Watershed (PDF)
    - 8th Grade Water Quantity: Water to Supply an Ever-growing Population (PDF)
    - First Grade Water Quantity Lesson: Here I Go 'Round My Watershed! (PDF)
    - Water Quantity Letter (PDF)
    - Kindergarten Water Quantity Lesson: Drip! Drop! Water Does Not Stop! (PDF)
    - Second Grade Water Quantity Lesson: Now You See It – Now You Don't! (PDF)
    - Water Quantity: What to Do and How to Do It (PDF)
    - WQT/7-1: Water Use Chart (PDF)

MY Sound

- The Impact of Atmospheric Nitrogen Deposition on Long Island Sound (PDF)
- Alternative Strategies for Hypoxia Management: Creative Ideas to Complement Advanced Treatment (PDF)
- Fact Sheet #1: Hypoxia in Long Island Sound (PDF)
- Toxic Contamination in Long Island Sound (PDF)
- Nutrient Reduction: New Solutions to Old Problems (PDF)
- Pathogens (PDF)
- The Impact of Septic Systems on the Environment (PDF)
- Water Conservation and Marine Water Quality (PDF)
- Wastewater Treatment (PDF)
- Supporting the Sound (PDF)
- Floatable Debris (PDF)
- How Low Dissolved Oxygen Conditions Affect Marine Life in Long Island Sound(PDF)
- Puttting the Plan in Motion (PDF)

SunWise

- SunWise Monitor, November 1999 (PDF)
  SunWise Monitor, April 2000 (PDF)
  SunWise Monitor, April 2001 (PDF)
- Mission: SunWise – Activity Book (PDF)
  Mission: SunWise – Activity Book (Spanish) (PDF)
- Sun Safety for Kids: The SunWise School Program (PDF)|
  The SunWise School Program Guide (PDF)
- Mission: SunWise (PDF)
  Mission: SunWise (Spanish) (PDF)
- Summertime Safety: Keeping Kids Safe from Sun and Smog (PDF)
- Action Steps for Sun Protection (PDF)
- Sunscreen: The Burning Facts (PDF)
- The Sun, UV, and You: A Guide to SunWise Behavior (PDF)
- What Is the UV Index? (PDF)
- UV Radiation (PDF)
- Ozone Depletion (PDF)